NEW FRONTIERS OF SOCIAL POLICY

Social Dimensions of Climate Change

Equity and Vulnerability in a Warming World

Robin Mearns and Andrew Norton, Editors

THE WORLD BANK
Washington, DC

© 2010 The International Bank for Reconstruction and Development / The World Bank

1818 H Street NW
Washington DC 20433
Telephone: 202-473-1000
Internet: www.worldbank.org
E-mail: feedback@worldbank.org

All rights reserved

1 2 3 4 13 12 11 10

This volume is a product of the staff of the International Bank for Reconstruction and Development / The World Bank. The findings, interpretations, and conclusions expressed in this volume do not necessarily reflect the views of the Executive Directors of The World Bank or the governments they represent.

The World Bank does not guarantee the accuracy of the data included in this work. The boundaries, colors, denominations, and other information shown on any map in this work do not imply any judgment on the part of The World Bank concerning the legal status of any territory or the endorsement or acceptance of such boundaries.

ISBN 978-0-8213-7887-8
eISBN 978-0-8213-8142-7
DOI 10.1596/978-0-8213-7887-8

Cover photo: © Arne Hoel/World Bank

Cover design: Naylor Design, Inc.

Library of Congress Cataloging-in-Publication Data
Social dimensions of climate change: equity and vulnerability in a warming world / Robin Mearns and Andrew Norton, editors.
    p. cm.
Includes bibliographical references and index.
ISBN 978-0-8213-7887-8 — ISBN 978-0-8213-8142-7 (electronic)
1. Social ecology. 2. Social change. 3. Climatic changes—Social aspects. I. Mearns, Robin. II. Norton, Andrew.
HM861.S62 2009
304.2--dc22

2009035885

In many developing countries, the mixed record of state effectiveness, market imperfections, and persistent structural inequities has undermined the effectiveness of social policy. To overcome these constraints, social policy needs to move beyond conventional social service approaches toward development's goals of equitable opportunity and social justice. This series has been created to promote debate among the development community, policy makers, and academia, and to broaden understanding of social policy challenges in developing country contexts.

The books in the series are linked to the World Bank's Social Development Strategy. The strategy is aimed at empowering people by transforming institutions to make them more inclusive, responsive, and accountable. This involves the transformation of subjects and beneficiaries into citizens with rights and responsibilities. Themes in this series include equity and development, assets and livelihoods, citizenship and rights-based social policy, and the social dimensions of infrastructure and climate change.

Titles in the series:

CONTENTS

PART II. SOCIAL JUSTICE AND CLIMATE ACTION

FIGURES

TABLES

For too long, global warming has been viewed as tomorrow's problem. The overwhelming evidence now suggests that climate change exacerbates existing development challenges, further exposing the vulnerability of the poor, and pushing those living on the margins closer to the edge. For those people most at risk, climate change is a crisis today. The rights, interests, and needs of those affected must take center stage as an issue of global social justice.

This volume brings together the final versions of papers that were presented first at an international workshop on the social dimensions of climate change, held by the World Bank in Washington, DC, on March 5–6, 2008. Other chapters were commissioned for this volume. A full report of the workshop is available separately. Taken together, these publications aim to amplify the voices of those people who are most at risk from climate change and to establish the basis for a compelling research and policy agenda on the social dimensions of climate change.

Our goal at the World Bank is to help developing countries realize the promise of sustainable development through the progressive achievement of the Millennium Development Goals. Our vision is to contribute to an inclusive and sustainable globalization—to overcome poverty, enhance growth with care for the environment, and create individual opportunity and hope.

Climate change alters the context of this work and thus demands new approaches, policies, and tools to help developing countries meet the challenges of reconciling climate action with the development and growth agenda.

For many years, the World Bank has been a focal point for financing on climate change, and today it is a major lender for renewable energy and energy efficiency. We also are at the forefront of developing the carbon market. The newly adopted Climate Investment Funds and Strategic

Framework on Development and Climate Change enhance our capacity to facilitate demonstration, deployment, and transfer of low-carbon technologies; and they increase our focus on building climate resilience in vulnerable nations.

Building on the foundation laid in the March 2008 workshop, the World Bank's Social Development Department is now leading efforts to promote socially inclusive, climate-resilient policies and operations in client countries.

I am confident that the innovative global agenda described in this volume will contribute to a more holistic analysis of climate change impacts on human and social systems; increase our understanding of vulnerability; and strengthen our capacity to build social justice, accountability, and equity into climate policy.

Katherine Sierra
Vice President, Sustainable Development
The World Bank

Robin Mearns is a lead social development specialist in the World Bank's Social Development Department and team leader for the social dimensions of climate change. Prior to joining the World Bank in 1997, he held research and teaching positions at the Institute of Development Studies, University of Sussex; and at the International Institute for Environment and Development. With a doctorate in geography, Mearns has more than 20 years of research, policy, and operational experience in natural resource management, environment, and rural development. He has focused particularly on land and agrarian reform, pastoral risk management, and community-driven development in Africa and Asia.

Andrew Norton is a lead social development specialist in the World Bank's Social Development Department. He has a doctorate in social anthropology, based on fieldwork in rural Mali and previously held positions in the United Kingdom's Department for International Development and the Overseas Development Institute. He has worked extensively on a range of topics, including poverty analysis, social protection, social issues in climate change, human rights–based approaches to development, and the politics of the budget process.

Arun Agrawal teaches environmental politics in the School of Natural Resources and Environment, University of Michigan. His research focuses on how power works in communities and in their efforts to govern, on adaptation and climate change, and on poverty and rural social life.

Simon Anderson heads the Climate Change Group at the International Institute for Environment and Development. That group works mainly on adaptation and poverty issues in the context of the most vulnerable countries and arid lands. Anderson comes from an agroecosystems background and is now developing an action-research portfolio on climate adaptation planning, equity, and effectiveness—looking particularly at the interface of endogenous and planned adaptation.

Halvard Buhaug is a senior researcher at the Center for the Study of Civil War at the International Peace Research Institute, Oslo, and he leads the Center's Working Group on Environmental Factors in Civil War. His research interests include the security implications of climate change and geographic aspects of armed conflict.

Justina Demetriades is a researcher at BRIDGE, a gender and development program at the Institute of Development Studies, University of Sussex. An experienced gender specialist, she is the manager of BRIDGE's Cutting Edge Program on Gender and Climate Change, and has written several articles on the topic.

Emily Esplen is a researcher at BRIDGE, a gender and development program at the Institute of Development Studies, University of Sussex. She has worked extensively on gender equality issues, including in its relation to climate change. She has a master's degree in the social anthropology of development from the School of Oriental and African Studies, London, and a bachelor's degree from the University of Cambridge.

Nils Petter Gleditsch is a research professor at the International Peace Research Institute, Oslo; editor of the *Journal of Peace Research*; and a professor of political science at the Norwegian University of Science and Technology, Trondheim. He served as president of the International Studies Association in 2008–09.

Jeffrey Hatcher is a policy analyst at the Rights and Resources Initiative, and previously worked for the United Nations Food and Agriculture Organization's Land Tenure Service. He holds a master's degree in cooperation and development studies from the European School of Advanced Studies, Pavia and a bachelor's degree in international relations from the College of William and Mary.

Rasmus Heltberg is a senior economist at the World Bank; he works and publishes on adaptation to climate change, poverty and vulnerability, social protection, and household energy. Before joining the World Bank in 2002, he was an associate professor of development economics at the University of Copenhagen, Denmark.

Lisa Jordan is an assistant professor of geography and public health at Florida State University. She has a master's of science degree in applied economics and a doctorate in geography. Her research focuses on the intersections of environmental justice, public health, demography, and geospatial sciences.

Steen Lau Jorgensen is the sector director of human development in the Middle East and North Africa Region at the World Bank. Prior to that, he served for seven years as the director of the World Bank's Social Development Department, in its Sustainable Development Vice Presidency Unit. Jorgensen, a Danish national, has held various adviser and senior positions at the Bank since he joined the institution in 1985, working on sustainable, community, and human development programs and policies across most regions of the world. He holds a graduate degree in economics from the University of Aarhus, Denmark.

Arvind Khare is the director of finance and policy at the Rights and Resources Initiative, and chaired the External Advisory Group of the World Bank on Forest Sector Strategy Implementation from 2005 to 2006. He is a natural resource management specialist with more than 20 years of experience in the nonprofit, corporate, and public sectors.

Jakob Kronik is an independent consultant with 20 years of experience working on the topics of indigenous peoples and environmental and social change. He holds a PhD in sociotechnological and environmental planning.

In recent experience for the World Bank, Inter-American Development Bank, and the United Nations, Kronik has coauthored a book on the social impacts of climate change for indigenous peoples in Latin America and has served as team leader for Country Environmental Assessments and Global Environmental Facility programs.

Megan Liddle has worked as program manager with the Rights and Resources Initiative (RRI), where she contributed to analytical work on forest tenure, forest-based poverty alleviation, and community-based forest enterprise. Prior to joining RRI, she worked with Forest Trends, as part of the Communities and Markets Latin America team. She has a degree in science, technology, and international affairs from Georgetown University.

Augusta Molnar is the director of communities and markets at the Rights and Resources Initiative. She is an anthropologist specializing in natural resource management. Molnar has worked for more than 20 years in the international development arena, mainly with the World Bank, in Asia and Latin America, focusing on forestry, land tenure, gender, indigenous peoples' issues, and rural development.

John Morton is a professor of development anthropology and an associate research director at the Natural Resources Institute, University of Greenwich. He has a doctorate in social anthropology and more than 25 years' experience in research and consultancy for rural development—especially in social, institutional, and policy aspects of livestock and pastoralist development. He was a lead author for the 2007 fourth assessment report of the Intergovernmental Panel on Climate Change.

Caroline Moser is a professor of urban development and director, Global Urban Research Centre, at the University of Manchester. Previously, she held positions at the Brookings Institution, the World Bank, and the London School of Economics. Her research includes urban poverty and household vulnerability, gender and development, urban violence and insecurity, and, most recently, intergenerational asset accumulation and poverty reduction strategies and implications for international migration and climate change.

Clionadh Raleigh is a lecturer in the geography department at Trinity College, Dublin, and an external researcher at the International Peace Research Institute, Oslo. She is a political geographer and her focus is on conflict, governance, and the social consequences of climate change in sub-Saharan Africa. Her current work concerns conflict and drought patterns in the Sahel belt.

Jesse Ribot is an associate professor of geography and directs the Social Dimensions of Environmental Policy Initiative at the University of Illinois. Previously, he worked at the World Resources Institute and at the Massachusetts Institute of Technology. His research focuses on natural resources and local democracy, distributional equity, commodity chains, resource access, and vulnerability.

David Satterthwaite is a senior fellow at the International Institute for Environment and Development and editor of the international journal *Environment and Urbanization*. He served on the third and fourth assessments of the Intergovernmental Panel on Climate Change and, in 2004, was awarded the Volvo Environment Prize. Recent books include *Adapting Cities to Climate Change* (co-edited with Jane Bicknell and David Dodman), published by Earthscan in 2009.

Paul Bennett Siegel has a master's degree and a doctorate in agricultural economics and broad experience in agricultural, rural, and regional diversification and development, applying asset-based and social risk management approaches to different contexts and countries. After an academic career in the United States and Israel, Siegel has been a consultant to the World Bank for the past 20 years.

William Sunderlin is leading field researcher on demonstration activities for Reducing Emissions from Deforestation and Forest Degradation (REDD) in a global comparative project on REDD at the Center for International Forestry Research. He has a doctorate in development sociology and has done research on the underlying causes of deforestation, poverty and forests, and rights and tenure in the forest, among other topics.

Ole Magnus Theisen is a PhD student at the Norwegian University of Science and Technology and the Center for the Study of Civil War at the International Peace Research Institute, Oslo. He holds a master's degree in political science from the Norwegian University of Science and Technology. His main interests are resource scarcities, climate change, and intranational violent conflict.

Camilla Toulmin is the director of the International Institute for Environment and Development. An economist by training, she has worked mainly in Africa on agriculture, land, climate, and livelihoods. Her work has combined field research, policy analysis, and advocacy and seeks to understand how environmental, economic, and political change affect people's lives and how policy reform can bring real change on the ground.

Dorte Verner is a senior economist and leads the Social Implications of Climate Change Program in the Latin America and Caribbean Region of the World Bank. She has published extensively in the areas of poverty and rural and social development. Verner joined the World Bank in 1996, having worked previously at the Development Center of the Organisation for Economic Co-operation and Development, and as a researcher at European University Institute, Sorbonne, and the University of Aarhus, Denmark. She holds a PhD in macroeconomics from the European University Institute, Italy, and a postgraduate degree from the University of Aarhus.

Andy White is the coordinator of the Rights and Resources Initiative (RRI). Prior to joining RRI, he served as senior director of programs at Forest Trends and as a natural resource management specialist at the World Bank. He has a PhD in forest economics and a master of arts degree in anthropology from the University of Minnesota.

This volume is based on original research commissioned for an international workshop on the social dimensions of climate change, held by the World Bank's Social Development Department (SDV), in Washington, DC, on March 5–6, 2008. The workshop would not have been possible without the efforts of many people within the SDV. Significant contributions to workshop organization were provided by Greicy Amjadi, Carina Bachofen, Mitos Benedicto, Joyce Chinsen, Danielle Christophe, Rasmus Heltberg, Steen Jorgensen, Megumi Makisaka, Carmen Martinel, Nicolas Perrin, David Post, Navin Rai, and Salam Syed. Special thanks also are due to Kristalina Georgieva (then acting vice president, Sustainable Development Network, World Bank) for her active support and participation. Overall guidance was provided by Steen Jorgensen (then director, SDV) and Caroline Kende-Robb (then sector manager, SDV).

A full report of the workshop is available separately and includes a list of all workshop participants and details of speakers, discussants, and session chairs. Although we cannot single out all the people who made important contributions, we are particularly grateful to the keynote speakers—Rt. Hon. Kim Campbell (former prime minister of Canada and former secretary general of the Club of Madrid), Her Excellency Dunya Maumoon (deputy foreign minister, Republic of the Maldives), Dr. Ibrahim Mayaki (former prime minister of Niger and current executive director of Rural Hub, Africa), Professor Bob Watson (chief scientist, Department for Environment and Rural Affairs, Government of the United Kingdom), and Sheila Watt-Cloutier (former chair, Inuit Circumpolar Conference)—for setting the scene so powerfully. Many thanks also to Rebecca Adamson (president, First Peoples Worldwide), Esther Mwaura-Muiri (founder and director, GROOTS Kenya), and Atiq Rahman (executive director, Bangladesh Center for Advanced Studies) for their invaluable support.

The introductory chapter of this volume draws extensively on the workshop summary report, which, in turn, drew substantially on discussions among participants during the workshop. This summary report (available at http://www.worldbank.org/sdcc) was written by Robin Mearns and Caroline Kende-Robb, with substantial contributions from Carina Bachofen, Gernot Brodnig, Edward Cameron, Megumi Makisaka, and Andrew Norton (all of SDV). Numerous people provided valuable comments on earlier drafts of the report, including Nilufar Ahmad, Nina Bhatt, Maria Donoso Clark, Maitreyi Das, Andrea Liverani, Radhika Srinivasan, Dorte Verner, Per Wam, and Carolyn Winter (all of the World Bank), Hans Olav Ibrekk (Norwegian Agency for Development Cooperation), and Ellen Wratten (U.K. Department for International Development—DFID).

Most of the subsequent chapters in this volume are the final, significantly revised versions of papers first prepared for the 2008 workshop. They have benefited substantially from anonymous peer review by colleagues inside and outside the World Bank. Thanks are due to the following reviewers for their invaluable contributions to this process: Arun Agrawal, Nilufar Ahmad, Jon Barnett, Richard Black, Gernot Brodnig, Edward Cameron, Siri Eriksen, Abha Joshi-Ghani, Daniel Hoornweg, Jon Lindsay, Peter Little, Andrea Liverani, Megumi Makisaka, Robert McLeman, Nicolas Perrin, Jorge Uquillas, and Katherine Vincent. Thanks also are due to Joseph Hamilton for editorial assistance and to Mustafa Pajazetovic for invaluable assistance in preparing the manuscript for publication. The diligent oversight of Cyprian Fisiy (director, SDV) in the latter stages of preparation of this volume is greatly appreciated.

Chapter 2 was commissioned for this volume with financial support from the governments of Norway and Finland, as donors to the Trust Fund for Environmentally and Socially Sustainable Development (TFESSD). Chapters 3 and 4 are based on stocktaking studies commissioned for the 2008 workshop, under contract with the International Peace Research Institute, Oslo; and chapters 8 and 9 are based on similar stocktaking studies commissioned for the workshop, under contract with the International Institute for Environment and Development. Financial support from the DFID is acknowledged gratefully for these four commissioned studies, for original work that informed the preparation of chapter 10, and for the workshop itself. Gerard Howe, Emmeline Skinner, and Ellen Wratten, among others, guided DFID's important contribution to this work. Chapter 5 is a revised version of a paper that first appeared in the *IDS Bulletin*. Chapter 6 is based on research conducted for a separate volume on the social impacts of climate change in Latin America and the Caribbean, with financial support from the Government of Denmark. Chapters 7, 10, and 11 are based on research carried out, in part, under projects executed by the World Bank's Social Development Department with financial support from the TFESSD.

Robin Mearns
Lead Social Development Specialist
Social Development Department
The World Bank

Andrew Norton
Lead Social Development Specialist
Social Development Department
The World Bank

A/R	afforestation/reforestation
AIDS	acquired immune deficiency syndrome
CDM	Clean Development Mechanism
COP13	Conference of the Parties, Thirteenth Session
GDP	gross domestic product
GHG	greenhouse gas
IPCC	Intergovernmental Panel on Climate Change
MA	Millennium Ecosystem Assessment
MDGs	Millennium Development Goals
NGO	nongovernmental organization
OECD	Organisation for Economic Co-operation and Development
ODA	official development assistance
PES	payments and markets for ecosystem services
REDD	reducing emissions from deforestation and forest degradation
RRI	Rights and Resources Initiative
SFTF	State Failure Task Force
UNFCCC	United Nations Framework Convention on Climate Change

CHAPTER 1

Equity and Vulnerability in a Warming World: Introduction and Overview

Robin Mearns and Andrew Norton

Climate change is widely acknowledged as foremost among the formidable challenges facing the international community in the 21st century. It poses challenges to fundamental elements of our understanding of appropriate goals for social and economic policy, such as the connection of prosperity, growth, equity, and sustainable development.[1] Human-induced climate change is widely perceived as threatening the long-term resilience of societies and communities throughout the world. It also poses unprecedented challenges to systems of global governance responsible for both controlling the scale of the phenomenon and responding to its impacts. This volume seeks to establish an agenda for research and action built on an enhanced understanding of the relationship between climate change and the key social dimensions of vulnerability, social justice, and equity.

Unequivocal scientific evidence, marshaled by the United Nations' Intergovernmental Panel on Climate Change (IPCC) in its Fourth Assessment Report, shows that greenhouse-gas (GHG) emissions from human activity—particularly burning fossil fuels for energy—are changing the Earth's climate (IPCC 2007). The *Stern Review* examines the economics of this complex phenomenon, a detailed understanding of which is needed to underpin an effective global response (Stern 2006); whereas others have focused their attention on its politics (Giddens 2009). As a recent *Human Development Report* made clear, there are glaring inequities in the distribution of responsibility for the causes of global warming and the distribution of its impacts among the nations and peoples of the world (UNDP 2007). Poor people in developing countries bear the brunt of its impacts while contributing very little to its causes. However, the human and social dimensions of climate change have been woefully neglected in the global debate—at least, until recently (Commission on Climate Change and Development 2009; Global Humanitarian Forum 2009; Roberts and Parks 2007; UNDP 2007).

This volume brings together the final versions of a set of papers first presented in an international workshop on the social dimensions of climate change, convened by the World Bank's Social Development Department in March 2008 (World Bank 2008), with additional material developed over the following year. The workshop brought together community activists, government representatives, former heads of state, leaders of indigenous peoples, representatives of nongovernmental organizations, international researchers, and staff of the World Bank and other international development agencies to help shape a global agenda on the social dimensions of climate change and their implications for effective climate action. This introductory chapter provides an overview of the major themes addressed in the workshop and elaborated in this volume. It articulates the broad outlines of a research, policy, and operational agenda on the social dimensions of climate change—necessarily integrated with existing mandates in international development—that places those people who are most affected by climate change front and center in the framing of equitable solutions.

Viewing climate change through a social development lens leads us, at the outset, to couch the agenda in terms of social justice, at all levels from the global to the local. The causes and consequences of climate change are intertwined deeply with global patterns of inequality. Climate change acts as a multiplier of existing vulnerabilities in a warming and transforming world. It threatens to roll back the hard-earned gains in poverty reduction and progress toward maintaining the Millennium Development Goals that already have been achieved (UNDP 2007). The global injustice of a world in which responsibility for the causes of climate change is inversely proportional to the degree of vulnerability to its consequences calls for equity and social justice to be placed at the heart of a responsive agenda on climate policy and action.

Motivated by concerns with equity and vulnerability, the policy and action-oriented agenda on the social dimensions of climate change outlined in this volume entails a dual-track approach, giving equal emphasis to both aggressive mitigation and pro-poor adaptation. But it follows from the social justice perspective that the transition to a low-carbon growth path, at least in the near term, should be undertaken primarily by richer countries and in those sectors that account for the bulk of GHG emissions: energy, heavy industry, buildings, and transport systems. For developing countries, the social dimensions of the climate change agenda largely concern adaptation to changes that now not only are considered unavoidable, but are already being faced by vulnerable communities on the front lines of a changing climate. Most of the chapters in this volume therefore focus on adaptation, and on the ways effective and equitable responses can and need to be integrated with existing approaches to development. But there is one major exception. Agriculture and land-use changes in developing countries account for approximately a quarter of all GHG emissions. That fact means that, with appropriate policy support, crucial and highly cost-effective mitigation decisions potentially lie in the hands of millions of poor farmers, pastoralists, and other land managers. Therefore, crafting equitable policy responses to mitigation in the agriculture and forest sector is also positioned firmly on the social dimensions of climate change agenda.

A point of departure for this twin agenda is to deepen understanding of who is vulnerable to the consequences of climate change, where, how, and why. This understanding includes not only how climate change contributes to vulnerability, but also how climate policy and response measures may magnify the effects of many existing drivers of vulnerability. In the short term, the biggest impact on poor people may result less from the changing climate itself than from policies adopted to mitigate climate change. Managing the potentially adverse social consequences of climate policy and efforts to maximize the benefits of climate action for the poor is central to the agenda. These concerns point to aspects of the policy and action agenda that otherwise might be overlooked, notably highlighting the importance of governance, institutions, voice, and social accountability in climate action and response measures at multiple levels. These themes resonate strongly with core preoccupations in the existing social development agenda (World Bank 2005a), but their application in the context of climate policy and action also has profound implications for the practice of development.

A focus on those who are most vulnerable highlights the urgency of the international community's current challenges to reach a fair and equitable deal at the 15th Session of the Conference of the Parties, under the United Nations Framework Convention on Climate Change (UNFCCC) in Copenhagen in December 2009, and to follow through on political commitments thereafter. Such a deal needs to include stringent targets to achieve GHG emissions reductions that would avoid dangerous global warming, generally defined as more than a 2-degrees Celsius average increase in surface temperature above preindustrial levels (Hadley Centre 2005). Given the inertia in the Earth's climate system and the slow pace of progress in the international negotiations, many people now would argue that such an ambitious "guardrail" may no longer be feasible (Parry et al. 2008). But owing to sea-level rise, the very existence of many low-lying, small-island developing states is threatened even at this level of warming—not to mention the lives and livelihoods of millions more people living in the Arctic, in the world's drylands, and in coastal mega-cities in the developing world. It is argued that a guardrail approach to mitigation that would serve the interests of those societies and communities on the front lines of today's changing climate also serves the interests of the rest of the international community, for whom the impacts may not become apparent immediately.

One track of the twin agenda on the social dimensions of climate change concerns urgent and ambitious action on mitigation to avoid the unmanageable (SEG 2007). The vexing questions are who should reduce GHG emissions, by how much, and by when? What would a fair, equitable deal look like? The equity challenge in the mitigation context boils down to reaching agreement about the basis on which to assign entitlements to the Earth's constrained ecological space—the ultimate global commons problem.

The other track of the social dimensions of climate change agenda concerns adaptation to manage the unavoidable consequences of changes that already are taking place in the Earth's climate system. The IPCC's Fourth Assessment Report warned that, although many impacts can be avoided, reduced, or delayed by mitigation—and regardless of how effective such mitigation efforts are—adaptation will be necessary to address impacts resulting from the warming that already is unavoidable because of past emissions (IPCC 2007). As former U.S. Vice President Al Gore[2] is reported to have said, "I used to think adaptation subtracted from our efforts on prevention. But I've changed my mind. . . Poor countries are vulnerable and need our help" (*The Economist* 2008).

The IPCC's Fourth Assessment Report identified for the first time those systems, sectors, and regions most likely to be especially affected by climate change (IPCC 2007). The most vulnerable systems and sectors are ecosystems such as tundra, boreal forest, mountain, Mediterranean-type ecosystems, mangroves and salt marshes, coral reefs and sea ice biomes; low-lying coasts, because of the threat of sea-level rise; water resources in low-latitude regions, as a result of increases in rainfall and higher rates of evapotranspiration; agriculture in low-latitude regions, because of reduced water availability; and human health, especially in areas with low adaptive capacity.

The most vulnerable regions are the Arctic, because of high rates of projected warming on sensitive natural systems; Africa, especially the sub-Saharan region, because of low adaptive capacity and projected changes in rainfall; small islands, because of the high exposure of the population and infrastructure to the risk of sea-level rise and increased storm surges; and Asian mega-deltas such as the Ganges-Brahmaputra and Mekong, because of their large populations and high exposure to sea-level rise, storm surges, and river flooding.

But in all those contexts, and in many more—even where incomes and adaptive capacity are relatively high—certain groups of people can be particularly vulnerable, including the poor, women, young children, and the elderly. It is these structural and situational drivers of vulnerability that concern us in determining appropriate responses in support of adaptation to climate change, and that form a major focus of a number of the chapters in this volume. At the core of this agenda is the need to understand the different ways people are vulnerable to the consequences of climate change by virtue of their geographic locations, livelihood sources, asset holdings, and social positioning; and the need to understand the implications for the appropriate tailoring of operational and policy responses in developing countries.

Most of the chapters in this volume are revised versions of papers originally prepared for the Social Dimensions of Climate Change workshop in March 2008, and they aim to take stock of the existing state of knowledge in core areas of this agenda. They include the chapters on climate change and armed conflict (chapter 3), climate change and migration (chapter 4), local institutions and climate change adaptation (chapter 7), implications of climate change for dryland societies (chapter 8) and for pro-poor urban adaptation (chapter 9), social policies and climate change adaptation (chapter 10), and climate change and the forests agenda (chapter 11). Chapter 2, dealing with understandings of vulnerability and implications for climate policy, was commissioned for this volume. Two other chapters have been adapted from work published elsewhere. An earlier version of chapter 5, on the gender dimensions of poverty and climate change adaptation, appeared as Demetriades and Esplen (2008); and chapter 6, on indigenous knowledge and climate change in Latin America and the Caribbean, is based on Kronik and Verner (forthcoming).

The volume is organized as follows. This introductory chapter first sets the scene by framing climate change as an issue of social justice at multiple levels, and by highlighting equity and vulnerability as the central organizing themes of an agenda on the social dimensions of climate change. It then provides an overview of the major elements of the twin agenda on pro-poor adaptation and the social dimensions of mitigation, and it introduces the main themes and arguments running throughout the volume.

Part I, comprising chapters 2 through 6, focuses on deepening our understanding of vulnerability in the context of climate change. Chapter 2 leads off with a review of existing theories and frameworks for understanding vulnerability, drawing out implications for pro-poor climate policy. The more integrative frameworks highlight the role of assets and institutions in contributing to livelihood security, and the importance of power relationships under conditions of risk and uncertainty that influence vulnerability outcomes for particular groups of people. Understanding the multilayered causal structure of vulnerability then can assist in identifying entry points for pro-poor climate policy at multiple levels. Building on such analytical approaches, chapters 3 and 4, respectively, consider the implications of climate change for armed conflict and for migration. These are two aspects of the emerging debate and literature on complex social responses to climate change that most merit further research and both of them highlight the difficulty in isolating climate-related from nonclimate drivers of vulnerability. Those chapters are followed by a discussion of two of the most important social cleavages that characterize distinct forms of vulnerability to climate change and climate action: gender (chapter 5) and ethnicity or indigenous identity (chapter 6)—in the latter case, focusing on the role of indigenous knowledge in crafting climate response measures in the Latin American and Caribbean region.

Part II explores in more detail the implications of the social dimensions of climate change agenda for policy and action in a range of priority contexts. Chapter 7 highlights the important mediating role of local institutions in achieving more equitable, pro-poor outcomes from efforts to support adaptation to climate change. Chapter 8 examines the implications of climate change for agrarian societies living in dryland areas of the developing world, and chapter 9 does the same for those living in urban centers. In both chapters, the authors describe and explain the distinct forms of vulnerability to climate variability and change faced by people living in these contexts; and they outline the elements of an integrated agenda to reduce vulnerability. In both cases, they highlight the buffering role of assets and the importance of accountable and representative forms of governance for pro-poor, local adaptation efforts. Chapter 10 considers the role of social policy instruments in supporting pro-poor adaptation to climate change; and it argues for a focus on "no-regrets" options that integrate adaptation with existing development approaches, albeit with modifications to take better account of the ways in which climate variables interact with other drivers of vulnerability. Finally, chapter 11 turns to the implications of climate policy and action for forest areas and forest people. Here again, themes of equity, rights, and social accountability emerge strongly as determining whether the outcomes will manage to balance emissions reduction objectives with those of protecting and promoting local livelihoods.

Climate Change and Social Justice

Climate change is often described as the defining global social justice issue of our time. It raises equity considerations between generations because actions taken or not taken today will affect future generations. It also has powerful implications for intragenerational equity today, among nation-states and among individuals and groups within societies. Climate change reinforces a vision of a world that is highly polarized—between heavy GHG-emitting countries and resource-poor countries that will suffer the worst consequences. The geographic distribution of per capita GHG emissions and levels of social and/or agroeconomic vulnerability are virtually mirror images of one another; when viewed together, they bear a striking resemblance to the already uneven global distribution of wealth and well-being (Dow and Downing 2007; SEG 2007). In short, climate change threatens to compound existing patterns of international inequality.

However, it is important not to present too stylized a dichotomy between the rich and poor worlds. There are middle-income (for example, oil-producing) countries with per capita emissions equal to or higher than those of countries in the Organisation for Economic Co-operation and Development (OECD), and highly populous developing countries with sharply rising per capita emissions. Countries in both categories also include many millions of people who are highly vulnerable to the consequences of climate change. Although it may appear convenient to characterize climate change mitigation as primarily the responsibility of rich countries, and adaptation as the chief concern of poor countries, things are not quite so simple. Vulnerability to climate change, viewed first and foremost as a development challenge, cuts across national borders.

Fairness in Mitigation and Adaptation

Article 2 of the 1992 UNFCCC requires "stabilization of greenhouse-gas concentrations in the atmosphere at a level that would prevent dangerous anthropogenic interference with the climate system...allow ecosystems to adapt naturally, ensure food production, and allow sustainable economic development" (UNFCCC 1992, p. 4). This twin agenda on both mitigation and adaptation has been described aptly elsewhere as a question of "avoiding the unmanageable and managing the unavoidable" (SEG 2007). What constitutes "dangerous" in this context is a value judgment determined by sociopolitical processes and informed by constantly evolving scientific, technical, and socioeconomic information. But deepening understanding in recent years has moved in the direction of favoring more rather than less urgent action on mitigation; and a powerful economic as well as ethical case for doing so has emerged.

Taking strong mitigation action is both good economics and consistent with aspirations for growth and development in poor and rich countries (Stern 2006, 2009). Although many uncertainties remain, and assumptions on discounting and risk aversion strongly affect the results, aggregate estimates of the economic costs of the impacts of unmitigated climate change range from 5 percent to 7 percent (if market impacts alone are taken into account) and up to 20 percent of annual global GDP (including broader, nonmarket impacts), equivalent to approximately $11 trillion (based on world GDP of $55 trillion in 2007). The *Stern Review* contrasts these estimates with the expected costs of taking action to cut emissions. The expected costs of reducing emissions, consistent with a midrange stabilization trajectory, are reckoned to be on the order of 1–2 percent of GDP per year, even without taking into account such additional economic benefits as energy access, energy security, or air quality (Stern 2006, 2009).

In terms of tackling the causes of climate change—or mitigation—a social justice perspective emphasizes the need for an equitable sharing among nations of the responsibility for reducing GHG emissions, based on an acknowledgment of the highly unequal distribution of past, present, and projected future emissions among them. This is what lies behind the Kyoto Protocol's guiding doctrine of "common but differentiated responsibility." But uncertainties associated with the inertia in the Earth's climate system, and the political challenge of negotiating the sharing of emissions reductions among those countries chiefly responsible for having generated the exiting *stock* of GHGs in the atmosphere and those projected to contribute increasingly to future *flows* of emissions, combine to make this exceedingly difficult to bring about in practice. Although making the transition to a low-carbon economy will be necessary in low- and middle-income countries as well as in the developed world if dangerous climate change is to be averted—and so in the direct interest of developing countries—access to affordable energy for the poor is a prerequisite for poverty reduction and economic growth in the shorter term. There also are questions regarding the social sustainability of some low-carbon technologies, such as hydropower and first-generation biofuels. Taking all these issues into account, the immense political challenge being confronted in the ongoing negotiations under the UNFCCC is to decide who should reduce emissions, by how much, and by when.

A number of principled proposals have been advanced to help address this enormously complex political question, including the Greenhouse Development Rights Framework (Baer et al. 2008) and the Contraction and Convergence approach (GCI 2008). Fairness in the context of the mitigation agenda has usually been interpreted to mean apportioning among nation-states the responsibility for reducing GHG emissions, according to their respective contributions to total global emissions. In such proposals, emissions are normally calculated on a per capita basis to draw attention to the underlying inequities in historical and current patterns of global consumption and the varying levels of carbon intensity driving consumption. Frameworks for attributing emissions on a per capita or personal basis also differ according to whether GHGs are attributed to the site of production or of consumption of consumer goods—an issue that makes a considerable difference in developing countries' perceived contributions to global emissions. Chakravarty et al. (2009) offer a framework for allocating national targets for fossil fuel carbon dioxide emissions that is derived from a fairness principle based on the "common but differentiated responsibilities" of individuals rather than nations. All of these proposals acknowledge that developing countries need "headroom" to increase per capita GHG emissions, at least in the short to medium term, consistent with their development aspirations to reduce poverty, meet the basic needs of their citizens, and attain the Millennium Development Goals. It is argued that such "subsistence" emissions are entirely different from the "luxury" emissions associated with consumption by rich people in rich countries (Agarwal and Narain 1991). Although it is true that developing countries will need to join in efforts to reduce global GHG emissions over the longer term—and that most of the least-cost mitigation options lie in the developing world (World Bank 2009)—these proposals all acknowledge those countries' immediate need for breathing space to address basic human development priorities.

In terms of tackling the consequences of climate change—or adaptation—a social justice perspective emphasizes that those whose lives and livelihoods are most vulnerable to the consequences of climate change and who have contributed the least to its causes should receive preferential support. This perspective should be both an integral part of and additional to existing efforts to reduce poverty and attain the Millennium Development Goals. Adaptation measures include long-term planning for infrastructure (water storage, supply, and sanitation; building codes; transport) and land use (flood management, conservation); agricultural diversification, research, and extension (for example, research on drought-tolerant crop varieties); streamlining legislation to avoid maladaptation (such as removing perverse incentives caused by farm subsidies, skewed water pricing, or inappropriate regulatory frameworks for land-use planning); planning for ex ante disaster risk reduction and ex post disaster response and recovery; and social policy measures, including the development of social protection systems (with various forms of personal and asset insurance), adaptation of public health priorities, and support to populations with special needs (including migrants). Such measures need to be mainstreamed into sector and national economic planning, while recognizing the aspirations of local communities and enabling them to adapt. Current approaches to adaptation may be missing major opportunities to engage creatively through institutional partnerships to support the diverse ways that local people are already adapting to climate change autonomously (Agrawal, chapter 7 of this volume).

Many estimates of the total expected costs of adaptation in developing countries have been made recently, going beyond existing development commitments. They vary widely, owing to differences in methodological assumptions and data limitations, the time frame considered (from the present to 2030), uncertainties associated with climate projections at national and subnational levels, and the difficulty in undertaking cost-benefit analysis for low-probability but catastrophic damage events. As a result, current estimates (derived from analyses in the *Stern Review* and conducted by Oxfam, the United Nations Development Programme, UNFCCC, and the World Bank) range from $3 billion to $135 billion per year (Commission on Climate Change and Development 2009; Project Catalyst 2009; World Bank 2009).[3] These estimates matter because they are being used to inform the international negotiations on the successor treaty to the Kyoto

Protocol, which runs out in 2012. From a social justice perspective, such calculations will have an important bearing on the level of international support that will be agreed necessary to flow from developed to developing countries in support of their adaptation efforts when countries convene in Copenhagen in December 2009.

Particular challenges stem from assumptions made about the share of losses from natural disasters that may be attributed to climate change rather than to existing climate variability. In its most recent report, the Global Humanitarian Forum estimates that approximately 40 percent of weather-related disasters in 2005 may be attributed to climate change, rising to 50 percent in 2030 (Global Humanitarian Forum 2009, p. 86). This results in an estimate of economic losses from climate change *today* amounting to $125 billion per year—roughly equivalent to the flow of 2008 official development assistance (ODA) from developed to developing countries, as tracked by the OECD's Development Assistance Committee (Global Humanitarian Forum 2009, p. 18). More conservative estimates using alternative methodologies to generate plausible scenarios still project increases in humanitarian spending on climate-related disasters during the next 20 years of between $57.0 million (a 32 percent increase over current levels) and $2.7 billion (a 1,600 percent increase), depending on assumptions regarding the expected frequency and intensity of future extreme events (Webster et al. 2008, p. 25).

Although considerable uncertainty remains concerning the actual sums likely to be needed, we can say with conviction that measures to reduce the risks associated with natural disasters and with responding to and recovering from them form a critical aspect of the social dimensions of a climate change agenda; and they need to be better integrated with adaptation and development efforts. Indeed, natural disaster risk reduction and recovery in many developing countries is the logical entry point to this wider agenda because natural disasters already form a very real and immediate threat to millions of people.

A Rights-Based Approach to Climate Change

A persuasive case is beginning to be made that climate change also poses threats to the realization of human rights and that the international human rights architecture is relevant in addressing climate change (OHCHR 2009; Orellana 2009; ICHRP 2008; Oxfam 2008; Seymour 2008). Although a specific human right concerning the environment has not been elaborated in a binding international convention, the fundamental right to an environment capable of supporting human society and the full enjoyment of human rights is recognized, in varying formulations, in the constitutions of more than 100 states and directly or indirectly in several international instruments.

Climate change now is accepted as the most immediate and far-reaching threat to the environment. Consequently, there is a growing concern that global warming will affect the full enjoyment of accepted human rights, including the right to life, the right to take part in cultural life, the right to use and enjoy property, the right to an adequate standard of living, the right to food, and the right to the highest attainable standard of physical and mental health (table 1.1).

A rights-based approach holds considerable promise for injecting urgency and ambition into global climate action while safeguarding the most vulnerable people in society. By focusing on equity and social justice, a rights-based approach offers both a compelling moral and ethical argument for action and a more authoritative basis for advocacy. It also helps give voice to vulnerable groups because, by design, human rights focus on the most vulnerable people on the planet. Moreover, by drawing on a body of human rights conventions, shared international laws, principles, and values stretching back more than 60 years, a rights-based approach could harness well-established technical, policy, and legal instruments in new ways to address climate change. There is a great deal of scope, for example, to examine how upholding important procedural rights—including access to information, decision making, and justice—could help promote social inclusion and accountability in climate action.

Perhaps most important of all, a rights-based approach helps identify duties and obligations. Under international law, governments are required to respect, protect, and fulfill their human rights obligations. To respect and protect rights, states must refrain from interfering with people's enjoyment of their rights. They also must prevent people's rights from being violated by third parties (such as by individuals, companies, or other countries). To fulfill rights, states must take action to enable the full realization of people's rights. This could be interpreted as requiring states to focus their adaptation measures on the most vulnerable communities within their jurisdictions.

It is important that a rights-based approach deal with inequities between countries as well as impacts on rights within countries. Those people who are immediately vulnerable to climate change have contributed little to its causes. They also lack the adaptive capacity to deal with its consequences.

Table 1.1. Possible Human Rights Implications of Climate Change

Natural Impacts of Climate Change	Impacts on Human Systems	Rights Implicated	International Conventions
Temperature issues	Increased water insecurity	Life	United Nations Universal Declaration on Human Rights, 1945 (for example, Article 3: "everyone has the right to life, liberty and security of person")
Risks of extreme weather events	Increased health risks/fatalities	Health	
Threats to unique systems	Changes in livelihoods	Means of subsistence	
Changes in precipitation patterns and distribution of water	Effects on the wider economy	Adequate standard of living	International Covenant on Economic, Social and Cultural Rights, 1966 (for example, Article 12: "The State Parties . . . recognize the right of everyone to enjoyment of the highest attainable standard of physical and mental health")
Threats to biodiversity	Changes in agricultural productivity and food production	Self-determination	
	Threats to security/cohesion	Water	
Sea-level rises, flooding, and storm surges	Effects on human settlements, land, and property	Culture	
		Property	International Covenant on Civil and Political Rights, 1966 (for example, Article 1.2: "in no case may a people be deprived of its own means of subsistence")
Large-scale singularities	Migration	Adequate and secure housing	
	Political/public services	Education	
	Damage to vital infrastructure	Gender, children's, and indigenous rights	Optional Protocol to the International Covenant on Civil and Political Rights, 1976
	Cultural integrity		Convention on the Elimination of All Forms of Discrimination Against Women, 1979 (for example, Article 14: "State Parties will take into account the particular problems faced by rural women . . . ")
	Decline in natural systems services		
	Distribution of impacts (vulnerable people will suffer most)		
	Aggregate demands		Convention on the Rights of the Child, 1989 (for example, Article 6: "State Parties shall ensure to the maximum possible extent the survival and development of the child")

Source: World Bank 2008, p. 25.

As a result, rights-based approaches advocate for substantial additional resources in support of climate change adaptation, preferably on grant rather than merely concessional financing terms, beyond existing commitments. But a rights-based approach should not be viewed as an "end-of-pipe" instrument, coming into effect only when a right is violated, a victim wronged, and an abuser identified. The best approach to human rights is one that establishes processes to ensure that violations never take place. This is particularly important in the context of climate change because some of the projected impacts will be difficult, if not impossible, to remedy and redress. Highlighting the importance of adaptation, then, is no excuse for failing to act urgently and ambitiously on the issue of mitigation. A balanced approach is needed, paying equal attention to both.

This section has highlighted numerous ways in which climate change must be viewed through a social justice lens. Global inequality in patterns of consumption emerges clearly as an intrinsic feature of climate change as a human-induced phenomenon. It is as relevant to understanding the uneven distribution of responsibility for the causes of climate change as to the asymmetrical impacts of climate change. A moral and ethical imperative following from this analysis highlights the obligations of richer countries both to reduce emissions rapidly and to provide adaptation support to poor countries. Within poor countries, a social justice perspective highlights the need to give priority to the poorest and most vulnerable groups in adaptation support; and to pay careful attention to ensuring that vulnerable groups benefit from measures to reduce GHG emissions, rather than be left worse off because of them. Table 1.2 offers a way to apply a social justice filter in linking the characteristics of climate change to their implications for social policy and action. The examples given there are merely illustrative rather than exhaustive, and are elaborated in further detail in the chapters that follow.

Poor People First: Who Is Affected, and How?

At a broad level of generalization, poor people in developing countries tend to depend directly on climate-sensitive sectors such as agriculture, forestry, and fishing for their livelihoods; therefore, they are more exposed to the impacts of climate change than are people in the developed world. People living in developing countries are also generally closer to the margin of tolerance to changing precipitation patterns, increased climate variability,

Table 1.2. Climate Change, Social Justice, and Policy Implications

Climate Change Characteristics	Social Justice Perspective	Implications for Policy and Action
Greenhouse gas emissions correlate with wealth and growth.	• Responsibility for climate change lies primarily with richer people in richer countries.	• Need to build global solidarity and momentum for climate action. • Developed countries have an ethical obligation both to reduce emissions rapidly and to provide adaptation support to poor people in poor countries. • Climate mitigation should not constrain energy access for poor people, nor should it constrain the growth paths of poor countries.
Climate change impacts differ according to people's power, wealth, and level of dependency on natural resources.	• The brunt of climate change impacts is borne by poor people in poor countries. They should receive preferential access to adaptation support. • Women will be disproportionately affected by climate change because social exclusion has a strong gender base that increases vulnerability to climate change for women. • Indigenous people are among the poorest and most socially excluded people globally. They rely on ecosystems particularly prone to the effects of climate change, including polar regions, humid tropics, high mountains, small islands, coastal regions, and semiarid deserts.	• There is a compelling need to understand the social dimensions of vulnerability by examining the assets, knowledge, institutions, and relationships that different groups have to help them cope with external threats. People can be more or less vulnerable according to age, ethnicity, caste, gender roles, sources of livelihood, ability to access public support, or ability to migrate. • An understanding of social difference must be translated into guarantees that people's enjoyment of fundamental human rights will not be compromised by climate change impacts.
Climate change will worsen water stress in many parts of the world through changes in rainfall patterns, glacial and snow melt, and rising salinity in low-lying coastal areas.	• Poor people will be most severely affected because they have less capacity to extract and store water. • Women in numerous contexts will see an increase in their labor burdens because they have primary responsibility for collecting water in many parts of the world.	• Investments in water resources must take account of the specific needs of poor people, particularly women, and build on local people's knowledge and priorities.

(continued)

Table 1.2. Climate Change, Social Justice, and Policy Implications (*continued*)

Climate Change Characteristics	Social Justice Perspective	Implications for Policy and Action
Extreme weather events (cyclones, storm surges) will become more frequent, with serious implications particularly for coastal areas.	• Poor people tend to be more vulnerable to injury, death, and destitution as a result of extreme weather events. For example, urban poor in informal slum settlements live in less-robust structures, tend to be unprotected by heavy infrastructural defenses, may be invisible to municipal authorities, and lack access to information. • Women frequently are more vulnerable to death and injury from cyclones and extreme weather events (owing to behavioral restrictions on mobility, restrictive dress codes, and lack of information).	• A range of actions is needed to empower the vulnerable and strengthen their resilience in the face of threats from extreme weather events, including • enhanced information to enable good choices about location and movement in the face of weather threats • enhanced tenure rights for housing to provide incentives to strengthen structures • enhanced rights for women to ensure that they have access to information and skills that will aid survival.
Carbon assets (trees, peat marshlands, rangelands, and the like) will increasingly be valued for their carbon sequestration properties in the struggle to contain and mitigate climate change.	• Poor people's rights in carbon assets—whether ownership or use and access rights—are critical to dignity and livelihood.	• Robust and accountable policy and institutional frameworks must be established to protect poor peoples' rights in carbon assets and to maximize the income streams they can derive from those assets.
Developed countries will increasingly seek to mitigate climate change through technological innovation.	• Technological innovations and shifts in market incentives in the North can have rapid and sweeping effects on the livelihoods of vulnerable people in the South (for example, the move to biofuels and resulting upward pressure on global food prices).	• Mitigation measures should be robustly analyzed ex ante to ensure that they do not cause damage to vulnerable people's livelihoods. • Rather than being regarded as a one-way transfer from North to South, technological innovation should capitalize on local people's knowledge.

Source: Authors' compilation.

and extreme weather events than are those living in developed countries, and thus more vulnerable to their effects.

These factors contribute to the challenges of livelihood security facing many of the world's most vulnerable people. These people include pastoralists and agropastoralists living in the world's drylands, exposed to increasing climate variability and changing means of temperature and precipitation. The inhabitants of low-lying, small-island developing states are highly exposed to sea-level rise and to the effects of coral bleaching on their economically important fishery and tourism sectors. Fishers and fish processors in coastal zones are among the most vulnerable groups in many developing countries. They face the loss of their major source of protein, and frequently have the lowest levels of human capital, the least transferable skill mix, and little or no access to land that could provide an alternative or secondary livelihood source. Poor people living in highly populous cities in low-lying deltas and coastal zones of developing countries are exposed both to sea-level rise and to flooding from storm surges. And those living in rural communities and major cities downstream of high-altitude glaciers in the Himalayas, Andes, Hindu-Kush, and other high-mountain regions are exposed to the loss of glacially regulated water resources for agriculture and drinking water. But in all geographic settings that are highly exposed to climate hazards, people are differently vulnerable, whether as a result of their sources of livelihood, levels of income and asset holdings, social class, gender, age, ethnicity, caste, access to public support, or ability temporarily or permanently to migrate in search of economic opportunities.

Just as levels and forms of vulnerability to the effects of climate change vary, so too does the capacity for societies to adapt to the changes that they will face. The adaptive capacity of developing countries is generally constrained by the limited availability of technology, weak institutional capacity, low levels of education, and inadequate financial resources. Other factors—such as poor nutrition patterns and weak health infrastructure—further contribute to higher losses of human life in developing countries as a result of climate change. The ways in which formal and informal social institutions interact also is thought to have an important bearing on societal resilience to extreme weather events, and possibly to slower-onset changes in climate as well. Social policy supporting gender inclusion and freedom of civic association, for example, combined with freedom of access to information and justice, has been observed to improve environmental performance; and preliminary evidence suggests that similar relationships also hold in the case of mortality from extreme weather events (Foa 2009). Overall, more research is needed to explore the contributory role of social institutions to societal resilience in the face of climate change.

This section and several chapters in this volume offer illustrative examples of how social difference influences human experiences of climate change and levels of adaptive capacity. It highlights the circumstances facing indigenous peoples (chapter 6), gendered vulnerabilities to the consequences of climate change (chapter 5), and the distinct forms of vulnerability facing poor people living in urban centers (chapter 9) and those living in arid and semiarid rural areas (chapter 8) of developing countries. Numerous other forms of social difference could be considered as well. Such knowledge is needed to inform strategic planning for adaptation at all levels, from the global to the local.

Indigenous Peoples

Indigenous peoples account for just 5 percent of the world's population, but they protect and care for an estimated 22 percent of the Earth's surface, 80 percent of remaining biodiversity, and 90 percent of cultural diversity on the planet (World Bank 2008, p. 33). They are also among the poorest and most socially excluded people in the world. Indigenous people tend to inhabit ecosystems particularly prone to the effects of climate change, including polar regions, humid tropical forests, high mountains, small islands, coastal regions, and arid and semiarid lands. Owing to their heavy dependence on these ecosystems as a source of livelihood, they tend to be disproportionately affected by climate change.

However, it is important that indigenous people not be seen merely as victims. Rather, they should be recognized as repositories of traditional ecological knowledge passed down over generations—knowledge that has enormous potential to complement and enrich existing scientific knowledge of climate change (see chapter 6). Although not immune to destabilizing shocks and stresses, indigenous people have often evolved customary institutions, rules, and practices that help ensure a sustainable relationship between society and the land and natural resources they depend on so directly.

Despite possessing such tacit knowledge, indigenous people have often been excluded from discussion and debate around the science and impacts of climate change on ecological and human systems. Indeed, certain proposed mitigation measures have the potential inadvertently to undermine the customary rights to indigenous peoples' lands and natural resources, thereby further contributing to their vulnerability and social exclusion (see chapter 11). This injustice can be addressed only through the recognition of indigenous peoples' rights and customary land and resource tenure, and through their inclusion as key partners and decision makers in the design and implementation of mitigation and adaptation interventions at global, national, and local levels.

Among such mitigation measures currently being developed, for example, are approaches to reducing emissions from deforestation and forest degradation (REDD), including payments for carbon held in sustainably managed and conserved forests under mechanisms such as the World Bank's Forest Carbon Partnership Facility and the United Nations' REDD mechanism (Angelsen et al. 2009). Much more work is needed, however, to design fair and equitable governance arrangements for forest carbon trading schemes, including clarifying tenure, property, and carbon rights (Cotula and Mayers 2009; chapter 11). If they are to succeed, REDD initiatives must be viewed within the wider context of efforts to promote sustainable forest management, involving indigenous and other forest people as active stewards of their forest environment and removing barriers to transparent, inclusive, and accountable forest governance (TFD 2008).

People of Different Genders

Current discussions on climate change pay scant attention to the significant ways in which climate change impacts and adaptation practices are gendered, calling for gender-specific adaptation strategies and action (chapter 5). Economic disadvantage, limited access to resources, dependency on male family members, and lack of power in decision making are factors that commonly contribute to women's vulnerability. Social exclusion also tends to be strongly gendered in ways that increase vulnerability to climate change for women and girls. For example, women typically outnumber men by 14 to 1 among those dying from natural disasters (Araujo et al. 2007, p. 1) because there may be cultural and behavioral restrictions on women's mobility, including restrictive dress codes that may prove deadly during floods; and in many societies, skills that could be essential for survival—such as tree climbing and swimming—are taught only to boys. Socially ascribed roles and responsibilities of women, such as collecting water and fuel, frequently lead them to be more directly dependent on natural resources and hence disproportionately vulnerable to the effects of climate change.

Lack of assets, shelter, resources, and access to information tends to make women more vulnerable than men in the face of natural disasters.

They are often reluctant to go to safe shelters during disasters, for fear of losing their children and their household assets, or may not receive warning information transmitted to men in public spaces. Such constraints meant that women suffered most during the 1991 cyclone and flood in Bangladesh. The death rate among women aged 20–44 was 71 per 1,000, compared with 15 per 1,000 among men (Aguilar 2004, p. 1; also see chapter 2 of this volume). Such gender differences in death rates attributable to natural disasters have been linked directly to women's economic and social rights (Neumayer and Plümper 2007). Rural women are likely to become increasingly vulnerable to slow-onset climate changes, owing to their reliance on agriculture and natural resources as livelihood sources. They may be forced to migrate to urban centers, particularly to informal settlements where they may be exposed further to conflict, crime, and violence and where supportive social institutions may be lacking.

As with indigenous peoples, women are not simply victims in the face of climate change; they are powerful agents of change and active managers of common-pool and household resources, because of their "triple roles" in productive, reproductive, and community-managing activities. So far, however, women have not been granted equal opportunities to participate in decision making related to climate adaptation and mitigation policies at the international and national levels; and the issue of gender has been conspicuously absent from UNFCCC deliberations. The policy debate fails to take into account the practical and strategic needs of women. Harnessing women's leadership skills and experience in community revitalization and natural resource management should be a priority in designing and implementing climate change adaptation and risk-reduction strategies.

Urban Poor People

Urban centers in low- and middle-income countries concentrate a high proportion of those people who are most at risk from the effects of climate change. Poor people tend to live in the interstices of such urban landscapes—in informal settlements, on steep slopes, along riverbanks and transport corridors, and on floodplains. Their security of land tenure is often precarious, and they lack the resources to invest in more protective forms of shelter. They see their lives, assets, environmental quality, and future prosperity threatened by the increasing risk of storms, floods, landslides, heat waves, and droughts. Water, drainage, and energy supply systems either are absent or are unable to cope with the increasing strains being placed on them.

Extreme weather events are among the most immediate threats to the urban poor in the developing world, and climate action in such settings should be approached in this context. But urban inhabitants are differently vulnerable to these climatic and environmental hazards, depending on their assets and capabilities—which, in turn, are influenced strongly by income, age, and gender. Understanding these differences and the synergies with a poverty reduction agenda is key to tailoring approaches to pro-poor adaptation with the aim of strengthening and protecting assets and capabilities at individual, household, and community levels (chapter 9).

There are two main reasons why strengthening and protecting the assets and capabilities of individual people, households, and communities are of far greater importance in developing than in developed countries. First, there are limitations in the capacity of municipal governments to support adaptation through the provision of protective infrastructure and services to low-income populations. Second, many municipal governments in developing countries are unwilling to work with low-income groups, especially those living in informal settlements who often are viewed as illegal squatters. Overcoming these barriers and enabling poor communities and municipal governments to work in partnership with one another are key to effective climate action in such settings.

Public support to build adaptive capacity must be tailored to meet the needs of vulnerable groups during three distinct stages of any risk cycle: predisaster risk limitation, immediate postdisaster response, and longer-term recovery from a risk episode. Common elements across all three stages are the need to increase the capacity of communities to make demands on municipal government for public support in providing protective infrastructure and services, and the need to increase the capacity of those local governments to respond. Therefore, strengthening the asset base of households and communities is a key means of building more competent, accountable local governments. But the converse is also true: increasing demand for better governance is a key means of strengthening the assets and adaptive capacity of poor households and communities.

People in Rural Drylands

More than 2 billion people—90 percent in developing countries—live in rural drylands that are characterized by a high degree of climatic variability and are highly susceptible to climate change (see chapter 8 of this volume). Dryland communities are also among the world's poorest, fastest growing, and politically least well-represented populations. They include pastoralists, agropastoralists, and those primarily dependent on rain-fed agriculture. Climate variability has long been a fact of life for those people living in drylands; and, in many ways, they are well adapted to an unpredictably variable environment. Their livelihoods and social institutions tend to be inherently oriented toward climate adaptation through flexible, ex ante strategies to reduce vulnerability, such as herd mobility, livelihood diversification, household splitting, migration, and traditional mechanisms for managing the conflicts that result from competition for scarce resources.

In addition to flexible livelihood strategies in situ, the strategies pastoralists and agropastoralists adopt for coping with and adapting to climatic variability often range much further afield. For example, pastoralists in East Africa and the Horn of Africa export livestock and livestock products to the Persian Gulf states and throughout the Middle East region. Long-distance migration to urban centers is commonly practiced as well, on a seasonal or longer-term basis, including forms of household splitting involving household members maintaining permanent bases in both urban and rural areas and sometimes even across international borders through migratory diasporas. These phenomena have been observed among pastoral communities in Mongolia, West Africa, and elsewhere. As is true for the urban poor in developing countries, levels of adaptive capacity vary according to the assets and capabilities of individuals, households, and communities. Only those households with sufficient human capital resources can afford to split or maintain dual or multiple bases; and this, in turn, is related to age or stage in the life cycle of the household.

The current state of knowledge does not allow for accurate, downscaled predictions of how climate change in arid and semiarid lands is expected to translate into drying and warming trends at regional, country, and local levels. For example, although highly uncertain, it is possible that parts of West and East Africa and South Asia will see an *increase* in rainfall. But climate change threatens significantly to alter the degree of variability and the frequency with which highly unusual events occur, as shown by the unprecedented floods in pastoral areas of Ethiopia and northern Ghana in recent years. Such events could strain existing adaptive capacities beyond their ability to cope. Inhabitants of drylands themselves often stress that rainfall patterns have become less predictable in recent years. This unpredictability makes it hard for farmers to time planting and other key decisions in the yearly cycle.

It is well known that vulnerability to the vagaries of the climate in dry-land areas is often exacerbated by nonclimate stressors, such as insecurity of land and property rights, disease, and conflict. Many drylands lie along the peripheries of nation-states; and dryland communities commonly move across national borders in search of pasture, water, and trading opportunities. Land tenure and rights are usually unclear—in part by design, because a certain degree of flexibility helps facilitate the mobility that is essential in coping with climate variability, but also often hotly contested, suggesting limits to such managed flexibility. Such border regions are often also hot spots for armed conflict; and the weak political integration and representation of pastoral groups in their own states frequently leaves them unprotected and vulnerable to violence. Traditional forms of livestock raiding have long been practiced among pastoral groups, such as those in northern Kenya, southern Sudan and Ethiopia, and northeastern Uganda. These raids had a certain redistributive logic, and played an important role in the initiation rites of young adult males. But these forms of raiding have recently become embroiled in the wider geopolitics of this contentious region; and the wide availability of automatic weaponry as a result of armed conflicts has spilled over, resulting in unprecedented levels of mortality in new and increasingly predatory forms of livestock raiding (Hendrickson, Armon, and Mearns 1998). Increasing climate variability and change could interact with underlying phenomena such as these to produce previously unseen forms of human insecurity.

Although these challenges are formidable, evidence suggests that there are multiple entry points available for enhancing climate resilience in dryland communities. If applied in combination and with a sufficient level of ambition, efforts to enhance resilience could yield effective results (chapter 8). These efforts build on traditional, adaptive livelihood strategies, but they introduce new approaches geared more toward empowering pastoralists and others living in drylands to make demands on their governments. Such approaches include land policy supporting pastoral mobility, water resource management (for example, rainwater harvesting), community-driven development building on local institutions, scaling up of community-based adaptation, drought early-warning systems, public information campaigns, regional initiatives to support the mainstreaming of adaptation in national and local plans, and strengthening citizen engagement with wider policy processes. These complementary approaches attempt to redress the unequal balance of power for people in drylands, while giving them voice and building capacity to adapt to climate change.

Complex Social Responses to Climate Change

To date, there is relatively little evidence of responses to drought and natural disasters involving large-scale conflict or migration. But we cannot count on studies of the past as necessarily good bases for future planning in this respect because perceptions of the likely future frequency of extreme weather events, or of the agents (who or what) responsible for them, may change; and there may be stepped changes in the severity of extreme events themselves. Furthermore, in the medium to long term, historically unprecedented levels of sea-level rise may cause population displacements on a substantial scale. Any of these possibilities has implications for migration responses (whether distress, planned, anticipatory, or labor-related) and policy. Little is known about the implications of climate change for conflict and state fragility, and this too is a priority area for further research.

Conflict and Human Security

Many of today's most fragile and conflict-prone societies are within the group of countries expected to be most severely hit by adverse climate changes in the coming decades. Without effective action on mitigation, climate change could overstretch the adaptive capacities of many societies, resulting in destabilization and violence and posing new challenges to national and international security (WBGU 2008; Smith and Vivekananda 2007; chapter 3 of this volume). That is likely to be particularly true in weak and fragile states with poorly performing institutions and governance structures. Between and within countries, conflicts could be triggered over scarce resources, especially water and land; over the management of migration; or over compensation payments from richer to poorer countries for adaptation finance.

This is a newly emerging area of research, and many of these propositions remain conjectural. It is proving to be a formidable challenge to establish empirical support for the links between conflict and climate effects. A number of regional hot spots have been identified in which constellations of security risks are thought to be associated with climate change and could develop into crisis hot spots (WBGU 2008). Such risks include climate-induced degradation of freshwater resources, climate-induced increase in storm and flood disasters, climate-induced decline in food production, and environmentally induced migration.

Three particular manifestations of climate change are thought to have substantial security implications: (1) increasing scarcity and variability in the yields of renewable natural resources on which human livelihoods depend; (2) sea-level rise, which is believed to have the potential to trigger massive population displacement, albeit over an uncertain time frame; and (3) intensification of natural disasters that would affect societies' resource base, infrastructure, and settlement patterns. The first of these manifestations—increased resource scarcity—has received most attention in literature discussing the human security implications of climate change. However, the precise nature of the relationship between resource scarcity and armed conflict, if any, remains unclear; and questions regarding causality remain to be explored in depth.

Buhaug, Gleditsch, and Theisen (chapter 3) suggest that climate change may increase the risk of armed conflict only under certain conditions and in interaction with several sociopolitical factors. Five social effects of climate change have been suggested as crucial catalysts of organized violence. First, increasing scarcity of renewable resources may cause unemployment, loss of livelihood and economic activity, and decreasing state income. Second, increasing resource competition may move opportunistic elites to intensify social cleavages. Third, reduced state income may hinder delivery of public goods, reduce political legitimacy, and give rise to political challengers. Fourth, efforts to adjust to a changing climate may have unintended side effects that could spur tension and conflict. Finally, worsened environmental conditions may force people to migrate in large numbers, thereby increasing environmental, social, political, and economic stress in receiving areas.

Past research has found societal factors such as low national income, large and ethnically diverse populations, weak and inconsistent political institutions, unstable regional "neighborhoods," and a recent history of large-scale violence to correlate closely with the risk of armed conflict. Given the social effects associated with climate changes, future negative security impacts are likely to be found in countries and regions that already experience organized violence. The mechanisms by which the negative impacts may spiral require further investigation in specific country cases; and there should be a focus on the sociopolitical catalysts, local mechanisms, and low-level violence.

Buhaug, Gleditsch, and Theisen (chapter 3) pinpoint several areas of research on conflict and climate change as deserving high priority in the future. First is to examine in greater depth what may be the plausible catalysts of conflict, including the conditions under which natural disasters may contribute to conflict. Second, it is important to widen the definition of conflict to include nonstate conflicts, to explore the influence of climate in shaping the course and outcome of ongoing conflicts, and to acknowledge regional implications. Third, in terms of research methods, pluralistic approaches that combine research traditions are needed, including a combination of more disaggregated research designs, quantitative analysis, and comparative analysis of historically grounded case studies. Such approaches would more readily allow complex relationships to be tested in a systematic manner that would enable investigators to draw general propositions.

Migration

Human mobility is a complex social phenomenon, and it is not simple to identify and isolate causal factors. Exploring the links between migration and climate change is even more challenging. Climate change may play a role in migration decisions, but usually is not the primary driver (Warner et al. 2009; chapter 4 of this volume), except where mobility has long formed an important strategy in adaptation to climate variability (chapters 7 and 8). Vulnerability to climate change is as much shaped by underlying inequities in access to power and resources as by the climate stressors themselves, leading differently positioned social actors to make very different choices with respect to mobility. For example, natural disasters vary widely in their potential to instigate migration. Whereas labor migration commonly intensifies in response to climate hazards, distress migration patterns are shaped by asset holdings, social networks, and available aid. It is also important to note that migration in response to climate variability is typically internal and short term, and is generally unrelated to conflict risk. Environmentally induced migration thus emerges from this analysis as a development concern rather than a future security issue.

Broadly speaking, the major climate change trends of relevance to migration are likely to be increasing temperatures and reduced rainfall, leading to water stress, drought, and reduced growing seasons in tropical and subtropical drylands (for example, the Sahel); sea-level rise, increased frequency of storm surges, and increased intensity of tropical cyclones, leading to flooding in low-lying and coastal regions; and higher temperatures leading to a longer growing season in temperate regions (such as northern Europe and Siberia).

Measuring the potential impact of climate change on human mobility is challenging. The standard approach to linking climate change and migration is to identify areas affected by climate change, count the number of people living there, and use that number to estimate the number likely to be forced to leave. However, this method is unsatisfactory because many factors drive migration, and the effect of climate variability and shocks is difficult to isolate. An alternative approach is to identify existing migration patterns and examine how demographic trends and climate change may affect the drivers of these specific migrations. Second-round impacts also will affect migration. For example, coping strategies may erode poor people's long-term ability to maintain a livelihood (strategies such as pulling children out of school or reducing food consumption), thereby strengthening the incentives for labor migration.

In general, better data are required on internal migration and displacement. These data can be used to determine how disasters vary in their effects on the basis of differential development, and they would allow for an assessment of local resilience and adaptation programs. Local, contextual information is probably more reliable than are country-level data because vulnerability differs significantly across disaster-affected communities. More research should be undertaken in areas where livelihoods are fragile and the margin for disaster is exceedingly narrow.

If we combine projected demographic and climate trends with current migration patterns, we may expect a number of impacts. Climate change impacts are likely to be more substantial where "push" drivers of migration coincide with high vulnerability to climate change and low capacity to adapt. In such areas (such as the Sahel and the highlands of Ethiopia), the pressure to migrate is likely to increase. Internal and cross-border movements appear more likely to be frequent responses to climate change impacts than do long-distance international movements, because economic losses associated with climate change may prevent people from investing in overseas migration and force them to look for work locally. For example, international labor migration from the Sahel region actually fell during the droughts of the 1970s and 1980s because people lacked the resources to invest in the journey (World Bank 2008, p. 47). Conflict-driven migration may be exacerbated by climate change, particularly where the change exacerbates conflict over natural resources (for example, Darfur). Temporary, short-distance "distress" migration is likely to rise as a consequence of climate shocks (such as droughts in the Sahel; floods in the Volta, Okavango, and Niger deltas). However, the numbers affected may be lower if anticipatory migration occurs in response to increased climate vulnerability. Some migration streams driven by the "pull" of economic opportunity also may be affected by climate change, including reduced opportunities for seasonal work in Eastern Sudan or Central Ghana and increasing employment opportunities in agriculture outside Africa. Coastal or low-lying areas will be vulnerable to sea-level rise and increased flood hazards. This, combined with increased overcrowding in urban areas, carries a risk of secondary migration.

There is also a need to further develop approaches for the managed relocation of populations whose livelihoods and settlements may not be secure. In the near future, the number of people needing to be relocated will be quite minimal. However, a best-practice strategy should be designed to deal with the most difficult future situations—the cases of small-island developing states and urban coastal areas. To prevent significant out-migration, present strategies could include protecting coastal infrastructure and limiting building in fragile coastal areas. In addition, regional agreements to facilitate postdisaster recovery should be developed in advance of need.

Pro-Poor Climate Action

While reducing the GHG emissions that cause climate change (mitigation) and addressing its consequences (adaptation) are separate conceptually, in practice they are closely interrelated realms of engagement. There are potential synergies and trade-offs between them, many of which are only just coming into sharper focus in the international debate. Most notably, in the context of forests, agriculture, and land-use change, a number of mitigation actions carry high social risks that could make it harder rather than easier for poor and vulnerable groups to adapt to the consequences of climate change.

Social Impacts of Mitigation

Many of the measures being proposed to reduce GHG emissions threaten to undermine further the livelihoods of those who are most vulnerable to the impacts of climate change. The promotion of first-generation or ethanol-based biofuels, efforts to reduce emissions from deforestation and degradation by putting a price on carbon, and large-scale investments in hydropower generation may have unintended consequences, resulting in the expropriation of poor and vulnerable groups' landholdings. This is because those who are among the most vulnerable—women and indigenous peoples—often have the least-secure property rights over the land and natural resources on which they depend for a livelihood and the weakest voice in pressing their claims. Whereas the international community has been slow to acknowledge these unintended consequences of climate action, the practical challenges in realizing co-benefits for local communities are likely to be formidable.

In terms of tackling the causes of climate change, it is vital to strengthen the benefits and reduce the potential costs of mitigation actions to poor people and their livelihoods. Much more work is needed to develop more inclusive and socially accountable approaches to hydropower development, biofuels, and forest carbon finance initiatives in the context of the UNFCCC negotiations, although the broad outlines of such approaches are beginning to emerge in some areas.

On the forests agenda, for example, we need an approach that proactively addresses social risks ex ante, applying lessons from international experience in promoting sustainable forest management in ways that actively involve forest people and communities (TFD 2008; chapter 11). This approach should include clarifying and securing their tenure, property, and carbon rights; removing barriers to transparent, inclusive, and accountable forest governance; and investing in institutional and organizational capacity to enable them to participate as full partners in climate action and in sustainable forest management, more broadly. Preliminary evidence suggests that such approaches could be among the most cost-effective means of "sequestering" (or capturing and storing) carbon, while empowering and creating incentives for local communities to serve as active stewards of their forest environments.

At the same time, the application of social safeguard policies is required to ensure that the various actors in forest carbon finance initiatives are held to account and comply with international agreements, minimum standards, and good practices. Much discussion in the forests context currently focuses on applying the principle of requiring the "free, prior and informed consent" of indigenous peoples—the core principle enshrined in the recently adopted United Nations Declaration on the Rights of Indigenous Peoples.

Synergies between Mitigation and Adaptation

To tackle the consequences of climate change effectively and equitably, it is essential to build the adaptive capacity and resilience of vulnerable social groups and the institutions that support them. This capacity enhancement includes the capacity to organize at the local level and form institutional partnerships in support of adaptation (chapter 7), to voice priorities and make claims on public policy, and to mediate and resolve potential conflicts arising from competition for resources. The importance of such approaches to building adaptive capacity is highlighted in chapter 8, with reference to people living in drylands in the developing world, and in chapter 9, with reference to the urban poor. In the case of the forests agenda (chapter 11), empowering and building the capacity of forest people to voice their priorities and make claims for public support are argued to be among the best ways to reduce potential risks associated with mitigation. Such approaches help overcome the potential trade-offs between mitigation and adaptation, and instead shift climate action in the direction of realizing synergies between the two agendas.

Tackling adaptation also requires seeing beyond infrastructure in deciding on priorities in allocating resources for adaptation. It is widely acknowledged that there is substantial overlap or complementarity between a climate-change adaptation agenda and what may be considered the existing realm of pro-poor development (chapters 2 and 10). Many societies are not well adapted even to existing climatic conditions, including the challenges posed by existing levels of climate variability; and this suggests that there is an "adaptation deficit" that needs to be addressed even before turning to an agenda that explicitly addresses the need to adapt to future climate changes. Measures to close this adaptation deficit are needed to help households and countries gear up for the expected increase in future climate volatility. Although such efforts may look a lot like development as we know it, they imply much more than "business as usual": interventions will have to be designed in different ways to take account of changing risk patterns and longer time horizons for adaptation.

There is wide scope for "win-win" or "no-regrets" policies and programs that simultaneously help address existing forms of vulnerability and provide a foundation for adaptation to future climate change (Heltberg, Siegel, and Jorgensen 2009). In many respects, sound development is the best form of adaptation: strong and accountable institutions, effective delivery of education and health care services, integrated water resources management, pro-poor agricultural research and extension, good infrastructure and a diversified economy all contribute to societal resilience.

The role of social policy in helping realize such synergies in climate action is highlighted in chapter 10, with an emphasis on coherence of actions to achieve effective adaptation. Social safety nets, social funds, community-driven development, microfinance, weather-based index insurance, and other mechanisms for social protection will be critical for helping poor people adapt and for helping when adaptation fails.[4] Inclusive and responsive institutions are needed to ensure that the provision of critical services (health care, housing, education) can adapt to a changing situation. And when population movements accelerate in response to climate change, there will be challenges in ensuring that migrants can acquire security of person and livelihood in their new homes and communities.

The Development-Adaptation Continuum

It is helpful to conceive of approaches to climate change adaptation as forming a continuum (figure 1.1). At one end of the continuum are adaptation approaches that overlap substantially with those aspects of the existing development agenda. These approaches to climate change adaptation seek to reduce poverty by addressing the underlying drivers of vulnerability, whether climate related or not (chapter 2). This recognizes that climate-related stressors more often than not interact with other drivers of vulnerability, as clearly illustrated by the cases of people living in drylands (chapter 8) and in urban settlements in low-income countries (chapter 9). Such approaches seek to strengthen governance, policies, and institutions through approaches that include community-based natural resource management, community-driven development, and social protection programs, with a strong emphasis on empowerment, participatory planning processes, community involvement in decisions, access to information, and institutional capacity building.

Figure 1.1. Development-Adaptation Continuum

| Addressing vulnerability drivers | Building response capacity | Managing climate risk | Confronting climate change |

Vulnerability focus   Impacts focus

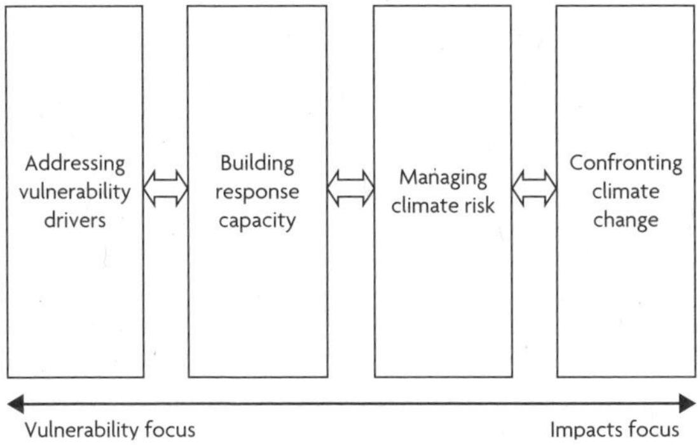

Source: Adapted from McGray et al. 2007.

At the other end of the spectrum are adaptation approaches that more explicitly address the direct impacts of climate change, such as improved weather data collection and forecasting capabilities. Between the two extremes are approaches that seek progressively to build climate response capacity and address climate risk management, including ex ante preparedness measures. These involve improved approaches to vulnerability and risk assessment; public services such as drought early-warning systems; improved intersectoral coordination (for example, for disaster risk preparedness and response); technical assistance (such as that designed to strengthen the capacity of health systems to address new diseases); public safety nets for those affected by disasters; and ex ante financing structures to pool risk at higher levels of aggregation, including global weather markets (for example, catastrophe bonds and index-based weather insurance). Within this range of action it is important that planning for social provision (including health, education, social protection, urban services, and regulation of labor markets) takes account of the increasing impacts of climate change.

However, many people would argue that the basis for providing adaptation finance under the UNFCCC is very different from the basis on which ODA is provided. Under the UNFCCC, structured according to the "Polluter Pays Principle," Annex 1 countries (mainly OECD countries, historically the chief emitters of GHGs) are obligated to provide adaptation support in the form of financial transfers and access to technology to non-Annex 1 countries (developing countries, in the main). Adaptation support, therefore, is regarded as a *right* of developing countries, in a way that goes beyond the moral and ethical foundations of ODA. This is why, in the context of the negotiations on a future global climate regime, it is vital that adaptation finance be seen as *additional* to existing flows of ODA (which developing countries fear it otherwise might displace).

Practically speaking, however—as the notion of the development-adaptation continuum makes clear—there is substantial overlap and complementarity in the types of activities that need to be supported by both ODA and adaptation finance. It is a necessarily integrated agenda. To make climate action work for, rather than against, the interests of poor and vulnerable people, policy and practical coherence is urgently needed between the realms of development and adaptation. To the extent that they may be addressed at times by different agencies and communities of practice on the ground, it also is essential that the lessons of decades of development experience are heeded and applied in operationalizing climate change adaptation.

Governance and Climate Action

A recurring theme throughout this volume concerns the governance of climate action at global, national, and local levels. Among the critical issues arising are which actors and institutions need to be involved, how to give voice to the vulnerable in crafting such governance arrangements, and how various forms of social accountability can be built in.[5] Governance and institutions powerfully shape adaptive capacity at the national level, and are critical in ensuring that results of mitigation efforts match intentions.

From community meetings to the corridors of the United Nations, the complex challenges involved in tackling the causes and consequences of climate change are fundamentally those of governance. First, the challenge of coordinating effective global action to mitigate climate change requires unprecedented collective action among nations, firms, and communities—constituting a massive challenge to systems of national and global governance. The barriers to the kind of collective action needed include the complex issues of equity and social justice discussed in this chapter, and issues of sequencing and market competitiveness. Second, the ability of governments to formulate effective and equitable adaptive responses depends on the strength and accountability of existing governance mechanisms. Third, societies' adaptive capacities are shaped in part by the support that good governance structures help facilitate. And, fourth, the distribution of costs and benefits within and among states in response to climate change fundamentally is determined by governance structures at the local, national, and international levels.

Although they often operate in isolation, governance mechanisms at the global, national, and local levels are vitally interlinked (chapters 2 and 7). For example, global frameworks shape national strategies, which influence local responses. Similarly, local knowledge can be drawn on to ensure that national and subnational policy responses are well aligned with local priorities (chapter 6); and the quality of subnational governance arrangements, including the character and density of civic institutions, can influence the success of national adaptation strategies (Foa 2009). This interconnectedness needs to be recognized so that policy makers effectively can assess the strengths and weaknesses of potential climate change interventions.

Voice and Representation

From the perspective of social sustainability, key elements of good governance are giving voice to and ensuring the representation of traditionally poor and marginalized groups. These are crucial when designing climate change adaptation strategies because these groups are likely to be most directly affected and already autonomously adapting to climate variability. Similarly, good governance is critical in the context of mitigation efforts, many of which may have adverse—if unintended—social consequences (chapter 11). Under these circumstances, good governance can be promoted through measures to increase voice and social accountability, maximize co-benefits to local communities, or empower communities by recognizing and strengthening their customary land tenure and other forms of rights so that they may benefit from the opportunities offered. Moreover, poor peoples' experiences may have significant value for decision makers who are designing these strategies at different levels.

Links between Global and Local Concerns

Although it undoubtedly is challenging, an important first step in examining how more effectively to link local and global concerns in a comprehensive governance agenda is to consider the following issues at the global, national, and community levels.

Issues at the Global Level. How can the voices of poor people and poor countries be taken into account in framing the terms of climate action? For example, the government of the Maldives regards the necessity to move out of one's homeland because of climate change as a *failure of adaptation* rather than an adaptive strategy. Similarly, who should draw the boundaries on the definition of "dangerous climate change"? Clearly, what is "dangerous" depends on one's current vantage point and how sensitive one's livelihood sources and strategies are to the projected impacts of climate change. We should be cognizant of the realities of vulnerable people in poor countries in the way we frame the boundaries of acceptable change.

Similarly, it has been observed that perspectives from developing countries of the South have tended to emphasize conceptions of "environmental justice" that differ from those of the North (Ikeme 2003). Specifically, a survey of the literature shows that the southern conception emphasizes three elements: *corrective/compensatory justice*, meaning the notion that the past must play a fundamental role in addressing present entitlements; *distributive justice*, implying that equal rights to GHG emissions should be accorded, in principle, to each individual in the world; and *procedural justice*, referring to the adoption of fair and inclusive procedures in the process of decision making on climate action. By contrast, much academic and policy work in the North focuses on determining the most economically efficient path for reducing emissions. Finding effective ways to advance effective climate action in a complex global economy will require that perspectives from poor people in poor countries become a significant part of the debate.

Issues at the National Level. In what ways can national climate policy objectives be designed to account for social and distributional issues? The policy debate will need to be informed by an understanding of the ways in which climate change will affect the livelihoods of a wide range of social groups—who, in turn, will need information on likely future effects to be able to voice their own priorities and concerns. To what extent may adaptation or mitigation strategies offer opportunities simultaneously to promote wider social development objectives? For example, a low-carbon urban design might include many provisions that would improve social conditions for the urban poor—including locating housing and businesses in such a way that journey times and costs are reduced, thus improving the conditions for walking and bicycling and improving public transport. How can technological solutions be designed and implemented in a way that is socially acceptable? How can governments ensure that the implementation of measures to protect forests, marshland, or rangeland as carbon sinks do not lead to a loss of property or use rights for poor populations? Transparent rules on ownership and institutions that are capable of applying them in a fair and proper way will be important factors in protecting the stakes of poor people in key natural resources in the era of climate action. How can we ensure that climate policy and response measures within developing countries are formulated in a fair and participatory manner? To strengthen voice and accountability in the development of national-level climate strategies and policies, it will be important to ensure that citizens have information about the choices and trade-offs involved, and that policy systems are responsive to their views.

Issues at the Community Level. How can communities be empowered to influence and guide adaptation strategies? The international development community can help in making information available at the community level. And it can work with governments to understand the ways in which communities are already adapting to climate change and variability, and the contributions that community-based adaptation will have to make in the future. To what extent do national policies empower local communities rather than make it harder for them to adapt? What government strategies have been successful in integrating community-based development approaches into national adaptation plans of action? Which types of strategies have undermined communities' capacities?

Social Learning and Adaptive Policy Making

Several distinct features of the global climate challenge weigh strongly in favor of ensuring that a socially inclusive learning process approach to adaptive policy making is placed at the center of climate action— particularly at the national, subnational, and local levels. The first features are the long time horizon over which decisions must be made, and the path-dependent nature of those decisions. The next feature is uncertainty: even if some changes are inevitable, their precise timing, location, and distributional impacts usually remain unclear. And third, effective and coherent climate change mitigation and adaptation involve coordinated action among a vast number of decentralized agents.

Adaptive policy making under these circumstances will require policy makers to treat policies and programs as ongoing experimental and learning processes, based on targets and milestones, strong performance-based monitoring and evaluation systems, and enabling frameworks for interactive engagement with multiple stakeholders. It also calls for much greater public participation in defining what climate change adaptation means in particular contexts. This could include, for example, the use of participatory scenario techniques with multiple stakeholders jointly projecting anticipated changes and planning for the kinds of public policy and other forms of support they need to help them adapt to those changes (Kuriakose, Bizikova, and Bachofen 2009).

Integrating Social Dimensions into Climate Policy

Viewing climate change through a social justice lens helps direct the future research and policy agenda toward priorities that most directly resonate with the people who are most vulnerable to the consequences of climate change. While the broad outline of a global agenda on the social dimensions is becoming clear, there is still much work to be done. This concluding section highlights some of the most important areas in which a social development agenda more effectively can inform climate action.

Advancing a "No-Regrets" Approach to Development

The international development community should advance a "no-regrets" approach that simultaneously promotes resilience to the adverse consequences of climate change and promotes sustainable development. For example, a number of social policy interventions (for example, social protection and insurance instruments) are "no-regrets" contributions to equitable risk management and potential springboards for local development. Indeed, even if some adverse effects of climate change do not emerge, investing in these "no-regrets" policies will leave countries developmentally better off than they otherwise would be. In short, we need to focus on identifying operational entry points that not only address the consequences of climate change, but also promote pro-poor growth.

Reframing the Issue

The social dimensions of climate change need to be more fully integrated into mainstream policy and planning within developing countries. Responsibility for action must be taken up at the appropriate level by the relevant agencies and government departments. In many developing countries faced with a laundry list of formidable development challenges, ensuring the protection of the natural environment may not be a major priority. Climate change still suffers from being viewed as primarily an environmental issue, and consequently it is relegated in many developing countries to the sole authority of generally weak environment ministries and agencies. Once it is reframed as a core challenge for sustainable development with powerful economic, social, and environmental implications, however, there is a greater likelihood of effective action by ministries of finance and planning, integration into national budgets, and take-up at the level of all relevant sector ministries and line agencies.

A major effort is required to raise awareness about, advocate for, and develop and share knowledge on the socioeconomic dimensions of climate change to ensure this wider sense of ownership over the agenda by powerful stakeholders in developing countries. The goal is to ensure that governments place equitable, socially just climate change responses at the heart of country-led poverty reduction, growth, and development strategies. Assisting governments in developing coherent and comprehensive climate policy involving effective interministerial and cross-sectoral coordination, and allocating sufficient funding for pro-poor adaptation and mitigation form major elements in achieving this goal.

Improving the Adaptive Capacity of Poor People

At the local level, the poor need to be informed of the risks posed by climate change and better equipped to deal with its impacts. People also should have access to, and be trained to use, social accountability tools and instruments (for citizen oversight and monitoring, as examples) so that they can hold government accountable for delivering results. The overall goal should be to help both governments and communities advance social development objectives through "no-regrets" policies and programs that also build resilience to climate change and promote good development.

Advancing an Emerging Policy Research Agenda

The realization that climate change is a core challenge for development has opened up space for a policy research agenda that focuses more on its economic and social impacts. From a social development perspective, a number of areas stand out as especially crucial. First, it is important to learn lessons from good practice in integrating local knowledge with scientific knowledge in formulating adaptation strategies. While local knowledge has much to offer in informing adaptation strategies, combining the two knowledge sources has proven challenging to date. Participatory mechanisms for bringing together local stakeholders' and expert knowledge, and then integrating them at scale, may be an important prerequisite, for example, for facilitating effective and culturally relevant action based on scientific forecasts and hazard warnings.

Second, it is important to develop indicators that can track progress toward achieving results on the social dimensions of climate change. Third, we need to make better use of frameworks and tools for social analysis (such as poverty and social impact analysis and participatory poverty and vulnerability assessments) when modeling the effects of climate change and assessing the impact that policies could have on the poorest and most vulnerable people. For example, we could include ethnographic research to understand how existing inequalities among groups and individuals may be reinforced or transformed under climate stresses, thereby shaping resource entitlements and well-being.

Finally, research should focus on making evidence-based policy recommendations for adaptation in the context of great uncertainty surrounding climate change. A number of these potential avenues for future research are outlined in the following chapters. Given that not all consequences of climate change play out in "real" time—that is, there often are significant lead-lag effects associated with threshold events—the research agenda must be much more focused on anticipating potential problems and using such tools as participatory scenario analysis to discern how people would respond to these challenges at the local level. There also is space to develop and apply innovative tools for multistakeholder engagement in devising action plans under conditions of uncertainty. For example, it would be instructive to bring together stakeholders from vulnerable communities with those from the private sector, governments, and civil society organizations jointly to develop robust scenarios for adaptation under alternative future scenarios of climate change impacts (Kuriakose, Bizikova, and Bachofen 2009).

Bringing Stakeholders Together for Greater Social Justice

Given that global agreements are difficult to achieve because of entrenched patterns of behavior in international relations, it is important to explore further the extent to which climate change creates opportunities to bridge global and local issues. Other important steps are to identify, assess, and engage the full range of actors with a stake in a socially just approach to climate change, including those from within governments, the private sector, civil society, and communities; and to identify potential areas for and new forms of partnership among these actors.

The key will be to find ways to create incentives for a diverse range of actors to speak with a common voice on the importance of addressing the social dimensions of climate change. Three types of incentives are emerging—financial, political, and moral. New financial and political incentives may help influence the extent to which a social justice approach to climate change—one emphasizing demand for voice, accountability, better governance, respect for rights, and acceptance of responsibilities—helps inform strategies to deal with climate change at local, national, and global levels. For example, public and consumer pressure on the private sector may provide an incentive to move toward more sustainable and socially responsible business models. Equally, the international climate negotiations must set an ambitious, long-term target for binding emissions reductions (with credible and enforceable intermediate milestones), if the carbon market is to thrive and encourage the private sector to change its business models. Last, but not least, if citizens are informed and empowered to demand more socially just and accountable forms of climate action, governments will come under pressure to respond, both from below and from their peers.

Notes

1. Jackson (2009), writing for the United Kingdom's Sustainable Development Commission, challenges the assumption of continued economic expansion in

rich countries, and proposes a model for economic management that prioritizes goals of equity and sustainability.

2. Gore and the IPCC jointly were awarded the Nobel Peace Prize in 2007.

3. The World Bank, supported by the governments of the Netherlands, Switzerland, and the United Kingdom, is currently conducting an empirical study on the economics of adaptation to climate change in developing countries. The study is intended to inform the climate change negotiations in Copenhagen in December 2009. It combines global estimates for key sectors with "bottom-up" analyses in seven countries, including vulnerability assessments and participatory scenario analysis with vulnerable groups in selected geographic "hot spots." See http://www.worldbank.org/eacc.

4. This is emphasized also by Stern (2009) and World Bank (2009).

5. "Social accountability" here refers to "the broad range of actions and mechanisms beyond voting that citizens can use to hold the state to account, as well as actions on the part of government, civil society, media and other societal actors that promote or facilitate these efforts" (World Bank 2005b, p. 4). It is particularly important that citizens be empowered to participate in defining appropriate forms of climate action on the part of their governments, and have access to information to enable them to monitor the consequences of these actions.

References

Agarwal, Anil, and Sunita Narain. 1991. *Global Warming in an Unequal World: A Case of Environmental Colonialism*. New Delhi, India: Centre for Science and Environment.

Aguilar, Lorena. 2004. "Climate Change and Disaster Mitigation." International Union for Conservation of Nature, Gland, Switzerland.

Angelsen, Arild, Sandra Brown, Cyril Loisel, Leo Peskett, Charlotte Streck, and Daniel Zarin. 2009. "Reducing Emissions from Deforestation and Forest Degradation (REDD): An Options Assessment Report." Report prepared for the Government of Norway. Meridian Institute, Washington, DC.

Araujo, Ariana, Andrea Quesada-Aguilar, with Lorena Aguilar, and Rebecca Pearl. 2007. "Gender Equality and Adaptation." Women's Environment and Development Organization and the International Union for the Conservation of Nature, Gland, Switzerland. http://www.genderandenvironment.org/admin/admin_biblioteca/documentos/Factsheet%20Adaptation.pdf.

Baer, Paul, Tom Athanasiou, Sivan Kartha, and Eric Kemp-Benedict. 2008. *The Greenhouse Development Rights Framework: The Right to Development in a Climate-Constrained World*. Revised 2nd ed. Berlin, Germany: Heinrich Böll Foundation, Christian Aid, EcoEquity, and the Stockholm Environment Institute.

Chakravarty, Shoibal, Ananth Chikkatur, Heleen de Coninck, Stephen Pacala, Robert Scolow, and Massimo Tavoni. 2009. "Sharing Global CO_2 Emission Reductions among One Billion High Emitters." *Proceedings of the National Academy of Sciences* 106 (29): 11884–88.

Commission on Climate Change and Development. 2009. *Closing the Gaps: Disaster Risk Reduction and Adaptation to Climate Change in Developing Countries.* Final Report of the Commission on Climate Change and Development. Stockholm, Sweden: Ministry of Foreign Affairs.

Cotula, Lorenzo, and James Mayers. 2009. *Tenure in REDD: Start-Point or Afterthought?* Natural Resources Issues No 15. London: International Institute for Environment and Development.

Demetriades, Justina, and Emily Esplen. 2008. "The Gender Dimensions of Poverty and Climate Change Adaptation." *IDS Bulletin* 39 (4): 24–31.

Dow, Kirstin, and Thomas Downing. 2007. *The Atlas of Climate Change: Mapping the World's Greatest Challenge.* 2nd ed. Berkeley, CA: University of California Press.

The Economist. 2008. "Climate Change and the Poor: Adapt or Die." September 13.

Foa, Roberto. 2009. "Social and Governance Dimensions of Climate Change: Implications for Policy." Background paper prepared for *World Development Report 2010: Development and Climate Change.* SDV Working Paper 115, Social Development Department, World Bank, Washington, DC.

GCI (Global Commons Institute). 2008. *Carbon Countdown: The Campaign for Contraction and Convergence.* London: GCI.

Giddens, Anthony. 2009. *The Politics of Climate Change.* Cambridge, U.K.: Polity Press.

Global Humanitarian Forum. 2009. *The Anatomy of a Silent Crisis.* Geneva, Switzerland: Global Humanitarian Forum.

Hadley Centre. 2005. *Avoiding Dangerous Climate Change: International Symposium on the Stabilization of Greenhouse Gas Concentrations.* Report of the International Scientific Steering Committee. Exeter, U.K.: Hadley Centre.

Heltberg, Rasmus, Paul Bennett Siegel, and Steen Lau Jorgensen. 2009. "Addressing Human Vulnerability to Climate Change: Toward a 'No-Regrets' Approach." *Global Environmental Change* 19: 89–99.

Hendrickson, Dylan, Jeremy Armon, and Robin Mearns. 1998. "The Changing Nature of Conflict and Famine Vulnerability: The Case of Livestock Raiding in Turkana District, Kenya." *Disasters* 22 (3): 185–99.

ICHRP (International Council on Human Rights Policy). 2008. *Climate Change and Human Rights: A Rough Guide.* Geneva, Switzerland: ICHRP.

Ikeme, Jekwu. 2003. "Equity, Environmental Justice and Sustainability: Incomplete Approaches in Climate Change Politics." *Global Environmental Change* 13: 195–206.

IPCC (Intergovernmental Panel on Climate Change). 2007. *Climate Change 2007: Synthesis Report: A Report of the Intergovernmental Panel on Climate Change.* Contribution of Working Groups I, II, and III to the Fourth Assessment Report of the Intergovernmental Panel on Climate Change. Geneva, Switzerland: IPCC.

Jackson, Tim. 2009. *Prosperity Without Growth? The Transition to a Sustainable Economy.* London: Sustainable Development Commission.

Kronik, Jakob, and Dorte Verner, eds. Forthcoming. *Indigenous Peoples and Climate Change in Latin America and the Caribbean.* Washington, DC: World Bank.

Kuriakose, Anne T., Livia Bizikova, and Carina Bachofen. 2009. "Assessing Vulnerability and Adaptive Capacity to Climate Risks: Methods for Investigation at Local and National Levels." Paper presented at the Seventh Open Meeting of the International Human Dimensions Programme on Global Environmental Change, Bonn, Germany, April 27–30. SDV Working Paper 116, Social Development Department, World Bank, Washington, DC.

McGray, Heather, Anne Hammill, and Rob Bradley, with E. Lisa Schipper and Jo-Ellen Parry. 2007. *Weathering the Storm: Options for Framing Adaptation and Development.* Washington, DC: World Resources Institute. http://pdf.wri.org/weathering_the_storm.pdf.

Neumayer, Eric, and Thomas Plümper. 2007. "The Gendered Nature of Natural Disasters: The Impact of Catastrophic Events on the Gender Gap in Life Expectancy, 1981–2002." *Annals of the Association of American Geographers* 97 (3): 551–66.

OHCHR (Office of the High Commissioner for Human Rights). 2009. "Report of the Office of the United Nations High Commissioner for Human Rights on the Relationship between Climate Change and Human Rights." Report No. A/HRC/10/61 to the Human Rights Council, Tenth Session, Geneva, Switzerland.

Orellana, Marcos. 2009. *Practical Approaches to Integrating Human Rights and Climate Change Law and Policy.* Washington, DC: Center for International Environmental Law.

Oxfam. 2008. "Climate Wrongs and Human Rights: Putting People at the Heart of Climate-Change Policy." Oxfam Briefing Paper No. 117, Oxfam International, Oxford, U.K.

Parry, Martin L., Jean Palutikof, Clair E. Hanson, and Jason Lowe. 2008. "Climate Policy: Squaring Up to Reality." *Nature Reports Climate Change* May 29. http://www.nature.com/climate.

Project Catalyst. 2009. "Adaptation to Climate Change: Potential Costs and Choices for a Global Agreement." Report of the Adaptation Working Group. ClimateWorks Foundation, San Francisco, CA.

Roberts, J. Timmons, and Bradley C. Parks. 2007. *A Climate of Injustice: Global Inequality, North-South Politics, and Climate Policy.* Cambridge, MA: MIT Press.

SEG (Scientific Expert Group on Climate Change). 2007. *Confronting Climate Change: Avoiding the Unmanageable and Managing the Unavoidable.* Report prepared for the United Nations Commission on Sustainable Development. Research Triangle, NC, and Washington, DC: Sigma Xi and the United Nations Foundation.

Seymour, Frances. 2008. *Forests, Climate Change and Human Rights: Managing Risk and Trade-Offs.* Bogor Barat, Indonesia: Center for International Forestry Research.

Smith, Dan, and Janani Vivekananda. 2007. *A Climate of Conflict: The Links Between Climate Change, Peace and War.* London: International Alert.

Stern, Nicholas. 2006. *The Economics of Climate Change: The Stern Review.* Cambridge, U.K.: Cambridge University Press. http://www.hm-treasury.gov.uk/stern_review_report.htm.

———. 2009. *The Global Deal: Climate Change and the Creation of a New Era of Progress and Prosperity.* New York: Public Affairs.

TFD (The Forests Dialogue). 2008. "Beyond REDD: The Role of Forests in Climate Change." TFD Secretariat, Yale University, New Haven, CT.

UNDP (United Nations Development Programme). 2007. *Human Development Report 2007/2008. Fighting Climate Change: Human Solidarity in a Divided World.* New York: UNDP.

UNFCCC (United Nations Framework Convention on Climate Change). 1992. Text of the Convention, as agreed at the United Nations Conference on Environment and Development, Rio de Janeiro, Brazil, June 3–14. http://unfccc.int/resource/docs/convkp/conveng.pdf.

Warner, Koko, Charles Ehrhart, Alex de Sherbinin, Susana Adamo, and Tricia Chai-Onn. 2009. *In Search of Shelter: Mapping the Effects of Climate Change on Human Migration and Displacement.* Geneva, Switzerland: CARE International.

WBGU (German Advisory Council on Global Change). 2008. *Climate Change as a Security Risk.* London: Earthscan.

Webster, Mackinnon, Justin Ginnetti, Peter Walker, Daniel Coppard, and Randolph Kent. 2008. *The Humanitarian Costs of Climate Change.* Medford, MA: Feinstein International Center, Tufts University.

World Bank. 2005a. *Empowering People by Transforming Institutions: Social Development in World Bank Operations.* Washington, DC: World Bank.

———. 2005b. "Social Accountability: What Does It Mean for the World Bank?" In *Social Accountability Sourcebook.* Washington, DC: World Bank. http://www.worldbank.org/socialaccountability_sourcebook/PrintVersions/Conceptual%2006.22.07.pdf.

————. 2008. "Social Dimensions of Climate Change: Workshop Report 2008." Social Development Department, World Bank, Washington, DC. http://www.worldbank.org/sdcc.

————. 2009. *World Development Report 2010: Development and Climate Change*. Washington, DC: World Bank.

PART I

UNDERSTANDING VULNERABILITY

TO CLIMATE CHANGE

Vulnerability Does Not Fall from the Sky: Toward Multiscale, Pro-Poor Climate Policy

Jesse Ribot

A society is ultimately judged by how it treats its weakest and most vulnerable members.

—Hubert H. Humphrey

If a free society cannot help the many who are poor, it cannot save the few who are rich.

—John F. Kennedy

If some combination of narcissistic morality and raw self-interest does not help reduce vulnerability, then perhaps some good analysis and political engagement may do so.

Analysis of vulnerabilities can help answer where and how society best can invest to reduce vulnerability. Analysis may not motivate all decision makers to make those investments, but it can give development professionals, activists, and affected populations fodder to promote or demand the rights and protections that can make everyone better off. Climate variations and changes present hazards to individuals and to society as a whole. The damages associated with storms, droughts, and slow climate changes are shaped by the social, political, and economic vulnerabilities of people and societies on the ground. Impacts associated with climate

The author thanks Arun Agrawal, Ashwini Chhatre, Floriane Clement, Roger Kasperson, Heather McGray, Robin Mearns, Andrew Norton, Ben Wisner, and several anonymous reviewers for their challenging constructive comments on drafts and in discussions to greatly improve this chapter, and Carina Bachofen for her diligent research assistance.

can be reduced through measures falling anywhere on a spectrum from climate change mitigation to reduction of the vulnerabilities of individuals and groups (McGray et al. 2007). This chapter calls for evaluation of the relatively neglected social and political-economic drivers of vulnerability at one end of this spectrum. The objective is to enable consideration of a full range of vulnerability-reducing policy responses. The chapter is concerned with the reduction of the everyday vulnerabilities of poor and marginal groups exposed to climate trends and events.

The world's poor people are disproportionately vulnerable to loss of livelihood and assets, dislocation, hunger, and famine in the face of climate variability and change (Cannon, Twigg, and Rowell 2003; and chapters 8 and 10 of this volume). Living with multiple risks, poor and marginalized groups must manage the costs and benefits of overlapping natural, social, political, and economic hazards (chapter 9). Their risk-minimizing strategies can diminish their incomes even before shocks arrive; and shocks can reinforce poverty by interrupting education, stunting children's physical development, destroying assets, forcing sale of productive capital, and deepening social differentiation from poor households' slower recovery (chapter 10). The poor also may experience threats and opportunities from development or climate action itself, such as efforts to reduce greenhouse-gas emissions in such sectors as household energy, land, and forest management (ICHRP 2008; O'Brien et al. 2007; Turner et al. 2003; chapter 11 of this volume).[1]

The good news is that policy can drastically reduce climate-related vulnerability. Although the best global data indicate human suffering and economic loss are worsening in the face of natural hazards,[2] the number of people affected compared with the total population is declining (Kasperson et al. 2005). This reduction in vulnerability is most pronounced in high-income countries, where higher levels of well-being, along with better infrastructure, policy, and planning, are successfully mediating the relationship between climate trends or events and outcomes. Effective climate action can further widen this gap between climate stressors and the risk of hardship.

In 1970, when Cyclone Bhola hit Bangladesh with 6-meter tidal surges, some 500,000 people perished (Frank and Husain 1971). In 1991, a similar storm, Cyclone Gorky, struck Bangladesh, causing 140,000 deaths. However, in 2007, when Cyclone Sidr (stronger than either Bhola or Gorky) hit Bangladesh with 10-meter tidal surges, fatalities were 3,406. Although population density increased in this area between the Bhola and Sidr catastrophes, the death toll was reduced dramatically (Government of Bangladesh 2008). Damage was reduced by Bangladesh's shift from a focus on

disaster relief and recovery to hazard identification, community preparedness, and integrated response efforts (CEDMHA 2007). Most important
were sophisticated early-warning and evacuation systems (Government
of Bangladesh, Ministry of Food and Disaster Management 2008; Batha
2007; Bern et al. 1993), which made Sidr 150 times less fatal than Bhola.[3]
This is an example of effective climate action.

Although there are notable policy successes, vulnerability of poor, marginalized, and underrepresented people remains widespread. In cases like
Bangladesh, women, the poor, and other marginalized groups are disproportionately and unacceptably vulnerable (Chowdhury et al. 1993). When
facing droughts in northeast Argentina, industry-dependent tobacco
growers are more vulnerable than independent agroecological farmers,
whose farms are more biodiverse, more technologically equipped, and less
exposed to external markets, and who have greater political negotiating
power (Kasperson et al. 2005). In Kenya, privatization of pasturelands has
improved security of some landholders, while making poorer and landless
people much more vulnerable (Smucker and Wisner 2008). In Northeast
Brazil, the poor remain vulnerable because of their dependence on rain-
fed agriculture combined with little access to climate-neutral employment
(Duarte et al. 2007). Poorer people excluded from access to services, social
networks, and land experience intensified climate-related vulnerabilities
and losses caused by unequal social relationships of power and representation. These kinds of problems are also a target for climate action.

The vast differences in damages associated with similar climate stressors in the same place at different times, from place to place, or among different social strata reflect the complex and nonlinear relationship between
climate and outcomes. The damages associated with climate events
result more from conditions on the ground than from climate variability
or change. Climate events or trends are transformed into differentiated
outcomes via social structure. The poor and wealthy, women and men,
young and old, and people of different social identities or political stripes
experience different risks while facing the same climate event (Blaikie
et al. 1994; Hart 1992; Agarwal 1990; Swift 1989; Watts 1987; Sen 1981;
Wisner 1976; chapters 5 and 9 of this volume). These different outcomes
are the result of place-based social and political-economic circumstances.
The inability to manage stresses does not fall from the sky. It is produced
by on-the-ground social inequality; unequal access to resources; poverty;
poor infrastructure; lack of representation; and inadequate systems of
social security, early warning, and planning. These factors translate climate vagaries into suffering and loss.

Poverty is the most salient of the conditions that shape climate-related vulnerability (Cannon, Twigg, and Rowell 2003; Prowse 2003; chapters 8 and 10 of this volume). The poor are least able to buffer themselves against and rebound from stress. They often live in unsafe flood- and drought-prone urban or rural environments; lack insurance to help them recover from losses; and have little influence to demand that their governments provide protective infrastructure, temporary relief, or reconstruction support (ICHRP 2008). Indeed, their everyday conditions are unacceptable even in the absence of climate stress. Climate stresses push these populations over an all-too-low threshold into an insecurity and poverty that violate their basic human rights (ICHRP 2008; Moser and Norton 2001).

Because the "adaptation" side of climate action aims to reduce human vulnerability, it cannot be limited to treating incremental effects from climate change so as to maintain or bring people back to their pre-change deprived state (also see chapter 10).[4] As Blaikie et al. (1994) point out, "despite the lethal reputation of earthquakes, epidemics, and famines, many more of the world's population have their lives shortened by unnoticed events, illnesses, and hunger that pass for normal existence in many parts of the world..." (p. 3; also see Kasperson et al. 2005 and Bohle 2001). It is this "normal" state that effective climate action must aim to eradicate if climate variation and change are to be downgraded from deadly threats to mere nuisances.

Following a brief review of vulnerability theory, this chapter frames an approach for analyzing the diverse causal structures of vulnerability and identifying policy responses that might reduce the vulnerability of poor and marginal populations. The chapter argues that an understanding of the multiscale causal structure of specific vulnerabilities—such as risk of dislocation or economic loss—and the practices that people use to manage these vulnerabilities can point to solutions and potential policy responses. Analysis of the causes of vulnerability can be used to identify the multiple scales at which solutions must be developed, and can identify the institutions at each scale responsible for producing and capable of reducing climate-related risks.

The chapters of this volume concur that there is insufficient knowledge on the social dimensions of vulnerability reduction intervention policies and programs.[5] This chapter outlines a policy-research agenda on causal structures of multiple vulnerabilities in different environmental and political-economic contexts so that causal variables can be aggregated to help develop higher-scale vulnerability reduction policies and strategies. The focus on causality builds on insight from successes of existing project approaches, such as social funds, social safety nets, or community-driven development (chapter 10), and successful adaptation support based on coping and risk-pooling practices (chapters 7 and 8). A focus on causal structure adds systematic attention to root causes at multiple scales. It identifies the proximate responses to risk, ordinarily conducted via projects and people's own coping arrangements, and attends to the more distant social, political, and economic root causes of vulnerability.

Vulnerability analysis and policy development are only first steps in a multistep iterative governance process. The chapter concludes with a discussion of governance, arguing that to tilt decision making in favor of the poor will require systematic representation of poor and marginal voices in climate decision-making processes.

Linking Climate and Society: Theories of Vulnerability

It is widely noted that vulnerability to environmental change does not exist in isolation from the wider political economy of resources use. Vulnerability is driven by inadvertent or deliberate human action that reinforces self-interest and the distribution of power, in addition to interacting with physical and ecological systems.

Vulnerability analysis often is polarized into risk-hazard and social constructivist frameworks (Füssel and Klein 2006; also see O'Brien et al. 2007 and Adger 2006). The risk-hazard model tends to evaluate the multiple outcomes (or "impacts") of a single climate event (see figure 2.1), whereas the social constructivist—or entitlements and livelihoods—approach characterizes the multiple causes of single outcomes (figure 2.2) (Adger 2006; Ribot et al. 1996; Ribot 1995). Integrative frameworks have grown mostly from the entitlements and livelihoods approach, but treat environment as a causal factor.

The two archetypal approaches ask different questions. The risk-hazard approach—which defines vulnerability as a "dose-response relation between an exogenous hazard to a system and its adverse effects" (Füssel and Klein 2006, p. 305)—is concerned with predicting the aftermath or "impact" of a given climate event or stress and with estimating the increment of damage caused by an intensification from "normal" climatic conditions to the conditions expected under climate change scenarios. Those who take this approach view people as vulnerable *to hazards*—locating risk in the hazard

Figure 2.1. Impact Analysis

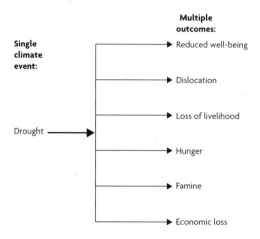

Source: Author's illustration.

Figure 2.2. Vulnerability Analysis

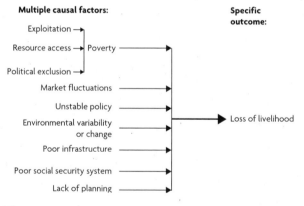

Source: Author's illustration.

itself. This approach usually is portrayed as inadequately incorporating social dimensions of risk (Adger 2006; also see Cannon 2000).

The social constructivists are asking what causes vulnerability. They consider people to be vulnerable *to undesirable outcomes.* They also are concerned with the likely aftermath of a climate event or trend. They view climate events and trends as external phenomena, and view the risk of disaster and suffering as social. Therefore, they place the burden of explanation of vulnerability within the social system. They locate risk within society. The entitlements and livelihoods approach is described as depicting "vulnerability as lack of entitlements" or a lack of sufficient means to protect or sustain oneself in the face of climate events where risk is shaped by society's provision of food, productive assets, and social protection arrangements (Adger 2006). The entitlements approach is often depicted as ignoring biophysical factors.

Integrative frameworks link these two views. These frameworks tend to be extensions of social constructivist models, rather than of risk-hazard approaches. Integrative frameworks view vulnerability as depending on both biophysical and human factors. One views vulnerability as having "an external dimension, which is represented...by the 'exposure' of a system to climate variations, as well as an internal dimension, which comprises its 'sensitivity' and its 'adaptive capacity' to these stressors" (Füssel and Klein 2006, p. 306). The Intergovernmental Panel on Climate Change views internal and external aspects as separate dimensions of vulnerability. These notions of internal and external aspects of vulnerability, however, are entirely contingent on how one draws the boundaries of the system under analysis.

Turner et al. (2003; also see Watts and Bohle 1993 and Blaikie 1985) have adopted an approach that avoids this boundary problem by tracing the causes of vulnerability from specific instances of risk—explaining why a given individual, household, group, nation, or region is at risk of a particular set of damages (see figure 2.2). By tracing causality out from each unit at risk, their model views the entire system as one integrated whole. Analyses of vulnerability then must account for all factors—biophysical and social—contributing to the stresses that affect the unit of concern (Kasperson et al. 2005). This causality-based integrative approach to vulnerability informs the available integrative analytic approaches described in the next section. It allows a multiscale, multifactor analysis of vulnerability.

Vulnerability Analysis

Two objectives of any vulnerability analysis for climate action are to identify who is vulnerable and to identify how to assist them. Analysts need to ask, *Where* should we spend public funds earmarked for climate adaptation, and *In what kinds of projects* should we invest in these places? The first question—how to target expenditures—requires identifying which regions (where), social groups (who), and things of value (what) are vulnerable. The question of what we need to invest in requires an understanding of the characteristics of the vulnerability of these people, places, and things and the reasons (why) they are at risk, so we can assess the full range of means for reducing that vulnerability. The questions *where, who,* and *what* are very different from *why*. Knowing where, who, and what tells us how to target expenditures. Knowing why tells us what to modify or improve in those targeted places and communities. *Why* also indicates the complexity and cost of short- and long-term solutions to vulnerabilities associated with climate variability and change.

Although impact assessments of the risk-hazard style can indicate that a place might be affected by a predicted climate change under given static, on-the-ground circumstances (a given level of exposure and ability to respond), they rarely tell us why the places and people or ecosystems are sensitive or lack resilience. Knowing likely impacts can help us target funding to particular places or to particular social groups or ecological systems. It cannot, however, tell us how to spend that money when we get there. Analysis of causes can help direct funds into vulnerability-reducing projects and policies. Climate action should be guided by both types of analysis. Much attention has been given to impact assessment, indicators, and mapping for targeting.[6] This section trains our attention on the elements of an analysis of causal structure of vulnerability.

The Causal Structure of Vulnerability

The two most common approaches to analyzing causes of vulnerability use the concepts of entitlements and livelihoods.[7] These approaches analyze the sensitivity and resilience of individual, household, or livelihood systems and, in some instances, the linked human-biophysical system. They tend to bring attention to the most vulnerable populations—the poor, women, and other marginalized groups. These approaches provide a starting point for analyzing the causes of climate-related vulnerability.

Entitlements and Livelihoods Approaches— Putting Vulnerabilities in Place

Sen (1981, 1984; also see Drèze and Sen 1989) laid the groundwork for analyzing causes of vulnerability to hunger and famine. Sen's analysis begins at the household level with what he calls "entitlements." Entitlements are the total set of rights and opportunities with which a household can command—or through which it is "entitled" to obtain—different bundles of commodities. For example, a household's food entitlement consists of the food that the household can command or obtain through production, exchange, or such extralegal legitimate conventions as reciprocal relationships or kinship obligations (Drèze and Sen 1989). A household may have an endowment or set of assets, including investments in productive assets; stores of food or cash; and claims they can make on other households, patrons, chiefs, government, or on the international community (Bebbington 1999; Drèze and Sen 1989; Swift 1989). Assets buffer people against food shortage. They may be stocks of food or things people can use to make or obtain food.[8] In turn, assets depend on the ability of the household to produce a surplus that it can store, invest in productive capacity and markets, and use in maintaining social relationships (Ribot and Peluso 2003; Berry 1993; Scott 1976).

Vulnerability in an entitlements framework is the risk that the household's alternative commodity bundles will fail to buffer them against hunger, famine, dislocation, or other losses. It is a relative measure of how prone the household is to crisis (Downing 1991; also see Watts and Bohle 1993, Downing 1992, and Chambers 1989). By identifying the components (that is, production, investments, stores, and claims) that enable households to maintain food consumption, this framework allows us to analyze the causes of food crises.[9] Understanding causes of hunger can shed light on policies to reduce vulnerability (Turner et al. 2003; Blaikie 1985). By analyzing chains of factors that produce household crises, a whole range of causes is revealed. This social model of how climate events might translate into food crisis replaces ecocentric models of natural hazards and environmental change (Watts 1983). By showing a range of causes, environmental stresses are located among other material and social conditions that shape household well-being. Hunger, for example, may occur during a drought because of privatization policies that limit pastoral mobility, making pastoralists dependent on precarious rain-fed agriculture (Smucker and Wisner 2008).

When environment (including climate) is located within a social framework, the environment may appear to become marginalized—set as one among many factors affecting and affected by production, reproduction, and development (see Brooks 2003). But this does not diminish the importance of environmental variability and change. Indeed, it strengthens environmental arguments by making it clear how important—in degree and manner—the quality of natural resources is to social well-being. These household-based social models also illustrate how important it is that assets match or can cope with or adjust to (that is, buffer against) these environmental variations and changes so that land-based production activities are not undermined by and do not undermine the natural resources on which they depend.[10] Leach, Mearns, and Scoones (1999) later called these environmental inputs to household sustenance "environmental entitlements" (also see Leach, Mearns, and Scoones 1997 and Leach and Mearns 1991).

"Environmental entitlements refer to alternative sets of utilities derived from environmental goods and services over which social actors have legitimate effective command and which are instrumental in achieving well-being" (Leach, Mearns, and Scoones 1999, p. 233). In that definition, these authors make four innovations. First, they expand Sen's concept of entitlements from an individual or household basis up to the scale of any social actors—individuals or groups. This enables analysis to be scaled to any relevant social unit (or exposure unit, in the case of climate-related analyses)—such as individuals, households, women, ethnic groups, organizations, communities, nations, or regions. Second, they introduce the notion of a subcomponent entitlement, a set of utilities that a particular resource or sector contributes to well-being—for example, environment.[11] Their third innovation also draws on Sen to show that "environmental entitlements enhance people's capabilities, which is what people can do or be with their entitlements" (p. 233). Last, they expand the idea of rights such that things may be "claimed" rather than just legally "owned." In this framing, claims may be contested—something Sen fails to capture. For example, when hunters close to Mkambati Nature Reserve in South Africa are banned from the reserve by state law, they continue hunting on the basis of customary rights that they view as legitimate. They claim their rights, contesting the state's claim (Leach, Mearns, and Scoones 1997). Hence, endowments such as natural resources that are not owned classically within a household still can be accessed through social relationships that may introduce cooperation, competition, or conflict mediated by systems of legitimization other than state law. With this insight, the authors introduce the notion that rights Sen takes as singular and static also may be plural (in the manner of Griffiths 1986 and von Benda-Beckmann 1981); and are based on multiple, potentially conflicting, social and political-economic relationships of access (in the manner of Ribot and Peluso 2003 and Blaikie 1985).

Watts and Bohle (1993) also place Drèze and Sen's (1989) analysis of household entitlements in a multiscale political economy. They argue that vulnerability is configured by the mutually constituted triad of entitle-ments, empowerment, and political economy. Here, empowerment is the ability to shape the higher-scale political economy that, in turn, shapes entitlements. For example, democracy or human rights frameworks can empower people to make claims for government accountability in provid-ing basic necessities and social securities (Moser and Norton 2001). Drèze and Sen have observed the role of certain types of political enfranchisement in reducing vulnerability—specifically, the role of media in creating crises of legitimacy in liberal democracies. Watts and Bohle go far beyond media-based politics to show that empowerment through enfranchisement puts a check on the inequities produced by ongoing political-economic processes. Although not outlined in their model, their approach indicates that direct representation, protest and resistance, social movement, union, and civil society pressures can shape policy and political processes or the broader political economy that shapes household entitlements (Ribot 1995). Moser and Norton view mobilization to claim basic rights as an important means for poor people to shape the larger political economy.

Multiple mechanisms link micro- and macro-political economies to shape household assets. Deere and de Janvry (1979) identify mechanisms by which the larger economy systematically drains income and assets from farm households. These mechanisms include tax in cash, kind, and labor (corvée); labor exploitation; and unequal terms of trade. These processes make people vulnerable because the wealth they produce from their land and labor is siphoned off—with the systematic support of social, eco-nomic, and environmental policies. For example, forestry laws and prac-tices in Senegal have prevented rural populations from holding onto profits from the lucrative charcoal trade (Larson and Ribot 2007), and foresters in Indonesia systematically extract labor from farmers and prevent them from trading forest products while allowing wealthy traders to profit (Peluso 1992). Scott (1976) also shows how peasant households are exploited in exchange for security. Peasants allow their patrons to take a large portion of their product or income in exchange for support during hard times.

Each household is affected by multiscale forces that shape their assets and well-being. Southern African farm households contend with climate variability, AIDS, conflict, poor governance, skewed resource access, and the erosion of their coping capacities. Although food production support is typical of food-security interventions, household-based research shows that food purchase supported by remittances and gifts is more important in enabling households to obtain food. Donors in the region supported climate early-warning systems, but these systems were found to do little to reduce vulnerability if not coupled with other measures. For example, farmers ask for guidance on specific actions to take, given forecast and warning information. Many farmers lack the capacity or resources (such as credit, surplus land, access to markets, or decision-making power) needed to turn climate information or specific guidance into action. These, then, are the proximate factors that shaped their vulnerabilities (Kasperson et al. 2005). The analyses framed by Watts and Bohle (1993), Deere and de Janvry (1979), and Scott (1976), as well as an analysis of the power and authority hierarchies in which households are embedded (Moser and Norton 2001), would give us insight into the larger political economy that would explain why credit is scarce and market access and representation are so limited.

Like entitlements analyses, livelihoods approaches (Cannon, Twigg, and Rowell 2003; Turner et al. 2003; Bebbington 1999; Blaikie et al. 1994) evaluate multiscale factors shaping people's assets. They build on entitlements approaches, but shift the locus of analysis from the household to multistranded livelihood strategies that also are embedded in the larger ecological and political-economic environment. They also shift attention from a focus on vulnerability to hunger toward an analysis of multiple vulnerabilities, such as risk of hunger, dislocation, and economic loss—a suite of factors closely related to the broader condition of poverty. In these approaches, vulnerability variables are connected with people's livelihoods, where a livelihood is "the command an individual, family or other social group has over an income and/or bundles of resources that can be used or exchanged to satisfy its needs. This may involve information, cultural knowledge, social networks, legal rights as well as tools, land, or other physical resources" (Blaikie et al. 1994, p. 9). Vulnerability in this framing is lower when livelihoods are "adequate and sustainable" (Cannon, Twigg, and Rowell 2003, p. 5). Livelihood models also explicitly link vulnerability to biophysical hazards by acknowledging that hazards change the resources available to a household and, therefore, can intensify some people's vulnerability (Blaikie et al. 1994).

In short, entitlements and livelihoods approaches form a strong basis for vulnerability analysis. They differ in the scale of the unit of concern and analysis (exposure unit) and the scope of factors that analysts view as impinging on that unit at risk—with livelihoods approaches being much broader. When taken together, they provide a powerful repertoire of analytic tools for vulnerability analysts. Both approaches (1) start with the unit at risk;

(2) focus on the avoidable damages it faces; (3) take the condition of the unit's assets to be the basis of its security and vulnerability; and then (4) analyze the causes of vulnerability in the local organization of production and exchange as well as in the larger physical, social, and political-economic environment. Vulnerability analysis differs greatly from the risk-hazard approaches that start with climate events and map out their consequences across a socially static landscape. Entitlements and livelihoods approaches put vulnerability in context on the ground, enabling us to explain why specific vulnerabilities occur at specific times in specific places.

Toward Pro-Poor Climate Action

Vulnerability to hunger, famine, and dislocation are correlated with poverty (Cannon, Twigg, and Rowell 2003; Prowse 2003; chapters 8 and 10 of this volume). Women, minorities, and other marginalized populations are also disproportionately vulnerable, sharing many of the vulnerabilities that poor people experience (chapter 5). For poor and marginalized populations, vulnerability reduction is poverty reduction and basic development (Cannon, Twigg, and Rowell 2003; also see Prowse 2003).

The weak within society tend to be of lower priority for those in power. Economically weak actors in urban slums or marginal groups far from the centers of power within semiarid or forested zones may be of little importance to people holding political office or involved in big business. They are likely to be low priority for governments, even in matters of disaster planning (ICHRP 2008; Blaikie et al. 1994). For instance, the extent to which slum dwellers are affected by extreme weather is a matter both of settlement location and of the level and quality of infrastructure and services such as water, sanitation, and drainage. These populations' lack of assets reduces their ability to adapt to changing conditions and prevents them from making political demands for investments to reduce their risk (chapter 9).

To counter biases against the poor and marginalized, vulnerability analyses and policies must be pointedly pro-poor. This section outlines an approach to pro-poor vulnerability analysis and a research agenda for identifying vulnerability reduction policies.

Pro-Poor Vulnerability Analysis

Entitlements and livelihoods approaches evaluate the causes of asset failure and of negative outcomes to identify means to counter the causes (Turner et al. 2003; Ribot et al. 1996; Ribot 1995; Watts and Bohle 1993; Downing 1991). This focus on negative outcomes favors poor and marginalized groups because they are overrepresented in at-risk populations. This tilt in favor of the poor also may be enhanced, of course, by analytic efforts that choose to study outcomes of most concern to the poor—outcomes such as hunger, dislocation or economic losses that push people over a threshold into poverty or extreme deprivation. The focus on causality can point toward solutions.

Coping and adaptation[12] studies identify vulnerability reduction strategies used by poor and marginalized populations and the means to support those strategies. Agrawal (chapter 7), for example, starts with household and community risk-pooling strategies and identifies institutions—civic, private, and public organizations—that support these strategies. His analysis gives insight into the roles of institutions (by which he means "organizations") and therefore into potential institutional channels for coping and adaptation support. Although this approach does not explain why people become vulnerable, it provides great insight into local-level vulnerability management and reduction.

Whereas analysis of coping or adaptation strategies can offer insight into causes of vulnerability, the entitlements and livelihoods approaches analyze the causal structure of vulnerability to identify a wider range of coping and adaptation opportunities (Yohe and Tol 2002; Mortimore and Adams 2001; Watts 1983; chapter 8). Coping approaches, as well as many project-based interventions, focus on means for adapting as well as on causes of adaptation and the ability to adapt. The vulnerability approach seeks to identify causes of the vulnerability—that is, causes of the risks to which people need to adapt.[13]

Tracing the causes of negative outcomes complements coping and adaptation approaches by enabling researchers and development professionals to conduct a full accounting of causality. Such a full accounting can indicate the policy options available for reducing vulnerability at its multiscale origins, rather than focusing only on coping or adapting in the face of hazards and stress (which tend to be responses to the most-proximate factors). For example, despite laws transferring forest management to elected rural councils in Senegal, foresters force councilors to give lucrative woodfuel production opportunities to powerful urban merchants—usually leaving the rural populations destitute (Larson and Ribot 2007). Forest villagers continue to rely on low-income rain-fed farming, and must cope with meager incomes. By focusing on the causes of destitution that put forest villagers on the margins, analysts might recommend means of policy enforcement rather than encouraging villagers to market other secondary forest products (as many projects are doing).

Vulnerability analysis most useful to policy makers starts from the outcomes we wish to avoid and works backward to the causal factors (Turner et al. 2003; also see Füssel 2007; Downing 1991; Blaikie 1985). In addition to favoring the poor, focusing on outcomes and their causes has other advantages: (1) it best matches policy to valued attributes of the system that we wish to protect; (2) it enables policy makers to place hazards as one variable among many affecting those attributes; (3) it brings attention to the many variables at multiple scales affecting valued attributes, steering analysts toward the many possible means for reducing the probability of negative outcomes or enhancing positive ones; (4) it enables comparative analysis of the many causes of negative outcomes, helping to focus policy attention on the causes that are most important, most amenable to reforms, and least costly to change—giving policy makers the biggest bang for their buck. Analyzing chains of causality (for example, Blaikie 1985)—by showing how outcomes are caused by proximate factors that, in turn, are shaped by more distant events and processes—can tell us what kinds of interventions might stem the production of vulnerability at what scales and, where relevant, who should pay the costs of vulnerability reduction.

Vulnerability reduction measures, of course, do not derive only from understanding causes. Indeed, some causes may be (or appear) immutable; others, transient, incidental, or no longer active. The objective of vulnerability analysis is to identify the active processes of vulnerability production and then to identify which are amenable to redress. Also identifiable are other interventions designed to counter conditions or symptoms of vulnerability without attending to their causes (such as support for coping strategies or targeted poverty reduction disaster relief). All forms of available analysis should be used to discover the most equitable and effective means of reducing vulnerability.

Identification of Multiscale Vulnerability Reduction Policies

Studies of coping strategies and lessons from successful development interventions provide valuable guidance for vulnerability reduction. Large-scale causes of vulnerability (such as unequal development practices), however, are less likely to receive attention in poverty reduction, vulnerability reduction, or adaptation programs. Identifying and matching solution sets or climate-related opportunities with responsive institutions at appropriate scales of social, environmental, and political-administrative organization provide an entry point into multiscale pro-poor climate action. Such action requires a systematic understanding of both proximate and distant dynamics that place people under stress or on the threshold of disaster. This section proposes a research agenda for identifying the range of causal factors shaping various vulnerabilities for groups at risk around the world, and a mapping of those causes onto solution sets for responsible and responsive institutions.

Different outcomes that we hope to avoid—such as loss of assets, livelihood, or life—are risks for different subgroups, and they have different associated causal structures. Different sectors will face different stresses and risks, and will have different response options (Parry et al. 2007). Within each case, the vulnerability of the poor (who have few resources to shield themselves or rebound from climate events and stresses) will be different from vulnerability of the rich (who are able to travel to safety and draw insurance to help them rebuild). Local, national, and international policies can be developed from an understanding of differences in the causal structures of vulnerabilities. Explaining differences will require an analysis of the multiple causal factors for a variety of vulnerabilities of concern. These causal data then must be aggregated to evaluate the best point of leverage for vulnerability reduction with respect to specific vulnerabilities and overall. Such an analysis should reveal the frequency and importance of different causes, pointing toward strategies to address the most salient and treatable causal factors.

Identifying the causal structure of vulnerability and potential policy responses can be a basis for developing a broad vulnerability reduction strategy. It involves aggregating causal structures over multiple cases of vulnerability among particular groups in particular areas to specific outcomes. This aggregation may have to be broken down by sector, ecozone, or hazard area to make the exercise manageable. The case studies also can serve as the basis for generating recommendations for local policy. More broadly, multiple case studies may help us comprehend the relative importance of different factors—both near and far—in producing and reducing vulnerability. These factors must be aggregated to identify the relevant scales and corresponding institutions for climate action. These steps set out a major research agenda for vulnerability reduction analysis. For this agenda to counter the biases against poorer populations, all of these steps must be consciously pro-poor. For example, the cases where such basic human rights as health, livelihood, and life are at risk must take priority over analysis of purely economic losses.

Indicators currently used to target poverty and vulnerability reduction interventions are a good starting point for identifying relevant study populations. Existing livelihoods approaches to vulnerability reduction already target the poor: strengthening their baseline nutrition, health, and morale; and addressing the underlying conditions of poverty, thus reinforcing their abilities to confront stressors and bounce back (Cannon, Twigg, and Rowell 2003). Vulnerability studies complement successful "self-help" and "social protection" coping and adaptation supports by indicating opportunities for higher-scale reforms (see chapter 10).

Thorough vulnerability analyses would indicate the need to reform the larger political economy of institutions, policies, social hierarchies, and practices that shape well-being, capacity for self-protection, and extended entitlements. For example, although social funds, community-driven development, and social safety nets are excellent means for responding to poor populations' immediate stresses and needs, examining causality through historical studies often reveals that the poverty these programs respond to is to the result of larger-scale, uneven development investment decisions and governance policies that limit the choices available to those affected by environmental disasters (chapters 4 and 10).

Vulnerabilities and their causes are diverse. Responses to vulnerability must be developed from detailed understandings of specific problems in specific places—general principles and models are insufficient. Case studies inform us of a particular set of dynamics and opportunities for vulnerability reduction in a particular place. It is from case studies that viable solutions can follow—both for specific places and more generally. To be complete, place-based approaches must take into account people's detailed knowledge of their social and production systems and the risks they face—experience with community-driven development teaches this lesson (Mansuri and Rao 2004). To make results of an analysis relevant and the implementation of recommendations feasible, investigations of vulnerability must consider local people's needs and aspirations and their knowledge of political-economic and social context in which any policy will have to be inscribed into law and translated into practice. Thus, although studies provide perspectives that communities may not be able to generate, the steps in developing a policy strategy for reducing vulnerability must be informed and open to influence by local citizens and their representatives.

Any vulnerability case study should include an evaluation of existing vulnerability reduction policy as well as a wide range of sectoral and regulatory policies (Burton et al. 2002). Existing policies deeply affect any given population at risk. Some policies are aimed at assisting them. Some may reduce vulnerability, and others help produce conditions of vulnerability. Policies, like institutions or organizations (as Agrawal suggests in chapter 7), can enable coping. They may also be systematically disabling (see Larson and Ribot 2007). Policies or their unequal implementation may selectively favor some actors and make others more vulnerable. Policies from all sectors have deep distributional implications. Coudouel and Paternostro (2005) and the World Bank's Poverty and Social Impact Analysis user's guide[14] suggest methods for analyzing the distributional effects of public policies. Such guidelines also can be applied to evaluating the vulnerability implications of policies and interventions.

When exploring the effects of policies and practices that shape vulnerability, or when analyzing potential vulnerability reduction measures, it is also important to account for a wide range of ancillary benefits (Burton et al. 2002). For example, in urban areas, asset building not only reduces immediate vulnerability, but also enables poor and middle-income people to make demands on their government for better services and infrastructure (chapter 9). Most adaptation measures will go far beyond reducing risk with respect to climate events. Hence, the set of benefits that follows from a given set of vulnerability reduction measures is also highly relevant in deciding the allocation of funds earmarked for development or for climate-related vulnerability.

Knowledge of problems and policy guidance can inform popular mobilization and policy making. Proposing policy solutions, however, is a small part of the political struggle for change. Calls for change must be backed by political voice and leverage. Bringing poor and marginalized groups into decision making through organizing or representation can reinforce their claims for justice, equity, and greater security in the face of a changing environment (Ribot 2004; Moser and Norton 2001).

Conclusion: From Climate Action Options to Institutions and Governance

Whereas vulnerability is always experienced locally, its causes and solutions occur at different social, geographic, and temporal scales. Identifying the causes of vulnerability points toward vulnerability reduction measures and the scales at which they best may be implemented. It also helps attribute responsibility to the polluters—providing a basis for compensation.[15] Vulnerability reduction or compensation policies are developed, promulgated, and implemented through institutions. So are the many other sectoral, economic, and social policies that have implications for vulnerability via their effects on resource access, market access, political voice, poverty, and economic distribution. Institutions also play numerous roles in supporting people's everyday coping and livelihood strategies (chapter 7). Systematically determining causes of vulnerability, identifying policy solutions, and mapping them to scales and appropriate institutions are three steps in a process that vulnerability reduction analysts and activists yet must conduct.

Institutions play several important roles in well-being and vulnerability. Leach, Mearns, and Scoones (1999) view institutions as mediating vulnerability by shaping access to resources (a part of endowment formation), the relationship between endowments and entitlements (rights and opportunities with which a household may command different commodity bundles), and the relationship between entitlements and capabilities (the range of things people may do or be with their entitlements). In their model, institutions enable people to obtain, transform, and exchange their endowments in ways that translate into contributions to well-being. As such, institutions support the needs of a plurality of subgroups, who can enter into competition and conflict when making claims to resources.

Agrawal (chapter 7) also emphasizes the role of institutions, showing how rural institutions structure risk and sensitivity in the face of climate hazards by enabling or disabling individual and collective action. Rural populations protect themselves by risk pooling via storage (over time), migrating (over space), sharing assets (among households), and diversifying (across assets). Exchange (via markets) can substitute for any of these risk-pooling responses. Rural institutions play a role in enabling each of these risk-reducing practices. In the 77 case studies Agrawal analyzes in his chapter, all of these practices depend on local institutions—mixes of public, civic, and private organizations.

Risk-pooling and exchange mechanisms constitute one set of practices that shapes vulnerability. Many other practices also produce or reduce climate-related vulnerabilities. Drèze and Sen (1989), for example, explore the role of media in influencing policy to prevent and respond to chronic hunger and famine. Leach, Mearns, and Scoones (1999) focus on the role of resource access, endowment formation, and entitlement mapping—the kinds of processes that might occasion the actors involved not to need to engage in risk pooling. Heltberg, Siegel, and Jorgensen (chapter 10) point to social protection interventions. Cannon, Twigg, and Rowell (2003) examine the role of networks (akin to Sen's 1981 extended entitlements);

Bebbington (1999) emphasizes social capital; Scott (1976) focuses on reciprocal relationships within a moral economy; Deere and de Janvry (1979) outline mechanisms by which economic gains are coerced or extracted from peasant households; and Moser and Norton (2001) emphasize the role of human rights and claim making.

Each of those enabling and disabling practices depends on different kinds of institutions—rules of the game and public, private, or civic organizations—at various scales. To map vulnerability-producing and vulnerability-reducing practices to institutional nodes for intervention, Agrawal's analytic approach to risk pooling could also be applied productively to each of these other practices. Each can be studied for its role in the causal structure of vulnerability. Each practice—whether reciprocity or social protection—depends on institutions that, when identified, can be targeted for reform or support. But attempting such interventions may generate social and political tension. As Leach, Mearns, and Scoones (1999) indicate, institutions and their networks may be in competition or conflict—some for enabling and others in support of disabling policies and practices.

The institutions responsible for and capable of responding to vulnerability are the loci of vulnerability governance. Governance (following World Bank 1992 and Leftwich 1994) is about the political-administrative, economic, and social organization of authority—its powers and accountabilities. It is about how power is exercised, and on whose behalf. As the global climate warms, decisions will be made at every level of social and political-administrative organization—from global conventions to the decisions of local governments, village chiefs, or nongovernmental organizations—to mitigate climate change, take advantage of its opportunities, and dampen associated negative consequences. Multiple decisions at multiple scales affect the livelihoods of urban and rural poor people. What principles of governance should guide decisions at each of these decision-making nodes? Who will decision-making bodies represent, and how? What distributions of decision-making powers and what structures of accountabilities will provide the most leverage for positive change and the checks and balances to protect poor urban and rural people's basic well-being and rights? These questions remain open.

Principles to govern climate action must be designed around the processes that shape vulnerability and the actors and organizations with authority and power to make decisions that can change these processes. The first step will be aggregating case-based analyses of causality, coping, and the role of institutions. That process can be tilted in favor of poor and marginalized populations by analyses that explain causes of asset and entitlement failure. To translate learning into action will be a long-term iterative process to negotiate the reshaping of policies and practice. All policies change distribution and, therefore, have advocates and meet resistance. Decision-making processes that are accountable and responsive to affected populations at least may tilt policies to favor the most vulnerable—because of their sheer numbers. Such a focus will promote the development of and engagement with representative decision-making bodies to ensure a modicum of influence by those people who are most in need.

For researchers, representation might mean incorporating the voice of local populations in their understanding of who is at risk, the problems at-risk groups face, and possible solutions, as well as sharing findings with affected populations and policy makers. For development professionals and policy makers, it will mean working with representative bodies and insisting that these bodies incorporate local needs and aspirations into the design of projects and policies. In global negotiations, it may mean requiring negotiators to engage in public discussions within their countries, or requiring national groups to organize and monitor their nation's negotiators. In local and national contexts, it may mean helping mobilize the poor and marginalized to make demands and to vote. Such governance practices may help avoid negative outcomes of climate action, and they could make climate action more legitimate and sustainable. Representing and responding to the needs of the most vulnerable populations might promote development that can widen the gap between climate and distress. Moving people away from the threshold of destitution by building their assets, livelihoods, and options will dampen their sensitivity, enhance their flexibility, and enable them to flourish in good times, sustain through stress, and rebuild after shocks.

Notes

1. For instance, this could occur if adaptations or mitigation efforts (such as reduced emissions from deforestation and decreased degradation) increase inequality within or among regions or social groups (O'Brien et al, 2007).
2. This trend holds, even without counting the 2004 Indian Ocean tsunami. Twice as many people were affected adversely by climate events in the 1990s as in the 1980s; and over the past four decades, major catastrophes have

quadrupled while economic losses have increased tenfold (Kasperson et al. 2005, pp. 151–52).

3. Like the storms in Bangladesh, Hurricane Katrina was a category 3 storm. Katrina's surge was 4 meters. Sidr was comparable to Katrina, which devastated New Orleans, Louisiana. But despite infamous Bush administration mismanagement, Katrina resulted in 1,300 fatalities (White House 2006).

4. The term "adaptation," although common in climate discussions, is highly problematic. It naturalizes the vulnerable populations; it implies that, like plants, they should adjust to stimuli. The term implicitly places the burden of change on the affected unit rather than on those causing vulnerability or bearing responsibility (for example, government) for helping with coping and enabling well-being. "Adaptation" also suggests "survival of the fittest," which is not a desirable ethic for society.

5. The U.S. National Research Council (Ramanathan, Justice, and Lemos 2007), the Intergovernmental Panel on Climate Change (Solomon et al. 2007), and the 2006 *Stern Review* all acknowledge the need for greater social science analysis.

6. On mapping and targeting, see Adger et al. (2004); Deressa, Hassan, and Ringler (2008); Downing (1991); and Kasperson et al. (2005).

7. For reviews of vulnerability approaches, see Adger (2006), Füssel and Klein (2006), and Kasperson et al. (2005).

8. According to Swift, "Assets create a buffer between production, exchange and consumption" (1989, p. 11).

9. An entitlements framework is very useful, but grossly incomplete—covering only a limited set of causes. For an analysis of its limits, see Gasper (1993).

10. Household models often are limited by their failure to account for intrahousehold dynamics of production and reproduction—but they do not have to be so limited. See, for example, Agarwal (1990), Carney (1988), Guyer (1981), Guyer and Peters (1987), Hart (1992), and Schroeder (1992).

11. This second innovation can be confusing because environmental claims in Sen's (1981) classic entitlements framework could be considered part of people's "rights and opportunities," and the alternative sets of utilities these can become would be part of the alternative commodity bundles people can command. Nevertheless, it is useful to view environment as contributing to people's endowments and alternative commodity bundles.

12. Coping is a temporary adjustment during difficult times, whereas adaptation is a permanent shift in activities to adjust to permanent change (Davies 1993; also see Yohe and Tol 2002).

13. Yohe and Tol (2002) focus on the determinants of adaptive capacity, but seek to identify causal structures rather than the causes of vulnerability.

14. The user's guide is available at http://siteresources.worldbank.org/INTPSIA/Resources/490023-1121114603600/12685_PSIAUsersGuide_Complete.pdf.

15. Füssel (2007) identifies three fundamental responses for reducing negative outcomes associated with climate change: mitigation, adaptation, and compensation. Mitigation assumes climate to be the major cause of problems. Adaptation and compensation require analysis of causality to identify a broader range of responsible factors and institutions.

References

Adger, W. Neil. 2006. "Vulnerability." *Global Environmental Change* 16 (3): 268–81.

Adger, W. Neil, Nick Brooks, Graham Bentham, Maureen Agnew, and Siri Eriksen. 2004. "New Indicators of Vulnerability and Adaptive Capacity." Technical Report 7, Tyndall Centre for Climate Change Research, University of East Anglia, Norwich, U.K.

Agarwal, Bina. 1990. "Social Security and the Family: Coping with Seasonality and Calamity in Rural India." *Journal of Peasant Studies* 17 (3): 341–412.

Batha, Emma. 2007. "Cyclone Sidr Would Have Killed 100,000 Not Long Ago." AlertNet. http://alertnet.org/db/blogs/19216/2007/10/16-165438-1.htm.

Bebbington, Anthony. 1999. "Capitals and Capabilities: A Framework for Analysing Peasant Viability, Rural Livelihoods and Poverty." *World Development* 27 (12): 2021–44.

Bern, C., J. Sniezek, G. M. Mathbor, M. S. Siddiqi, C. Ronsmans, A. M. Chowdhury, A. E. Choudhury, K. Islam, M. Bennish, E. Noji, and R. I. Glass. 1993. "Risk Factors for Mortality in the Bangladesh Cyclone of 1991." *Bulletin of the World Health Organization* 71 (1): 73–78. http://whqlibdoc.who.int/bulletin/1993/Vol71-No1/bulletin_1993_71(1)_73-78.pdf.

Berry, Sara S. 1993. *No Condition Is Permanent: The Social Dynamics of Agrarian Change in Sub-Saharan Africa*. Madison, WI: University of Wisconsin Press.

Blaikie, Piers. 1985. *The Political Economy of Soil Erosion in Developing Countries*. London: Longman Press.

Blaikie, Piers, Terry Cannon, Ian Davis, and Ben Wisner. 1994. *At Risk: Natural Hazards, People's Vulnerability and Disasters*. London: Routledge.

Bohle, Hans-Georg. 2001. "Vulnerability and Criticality: Perspectives from Social Geography." *IHDP Update* 2/01: 3–5.

Brooks, Nick. 2003. "Vulnerability, Risk and Adaptation: A Conceptual Framework." Working Paper 38, Tyndall Centre for Climate Change Research, University of East Anglia, Norwich, U.K.

Burton, Ian, Saleemul Huq, Bo Lim, and Emma Lisa Schipper. 2002. "From Impact Assessment to Adaptation Priorities: The Shaping of Adaptation Policy." *Climate Policy* 2: 145–49.

Cannon, Terry. 2000. "Vulnerability Analysis and Disasters." In *Floods*, ed. Dennis J. Parker, 43–55. London: Routledge.

Cannon, Terry, John Twigg, and Jennifer Rowell. 2003. "Social Vulnerability, Sustainable Livelihoods and Disasters." Report to the Department for International Development, Conflict and Humanitarian Assistance Department and Sustainable Livelihoods Support Office, London.

Carney, Judith. 1988. "Struggles over Land and Crops in an Irrigated Rice Scheme." In *Agriculture, Women and Land: The African Experience*, ed. Jean Davidson, 59–78. Boulder, CO: Westview Press.

CEDMHA (Center for Excellence in Disaster Management and Humanitarian Assistance). 2007. "Cyclone Sidr Update." November 15. http://www.coedmha.org/Bangladesh/Sidr11152007.htm.

Chambers, Robert. 1989. "Vulnerability, Coping and Policy." *IDS Bulletin* 20 (2): 1–7.

Chowdhury, A. Mushtaque R., Abbas U. Bhuiya, A. Yusuf Choudhury, and Rita Sen. 1993. "The Bangladesh Cyclone of 1991: Why So Many People Died." *Disasters* 17 (4): 291–304.

Coudouel, Aline, and Stefano Paternostro, eds. 2005. *Analyzing the Distributional Impact of Reforms: A Practitioner's Guide to Trade, Monetary and Exchange Rate Policy, Utility Provision, Agricultural Markets, Land Policy, and Education*. Washington, DC: World Bank.

Davies, Susanna. 1993. "Are Coping Strategies a Cop Out?" *IDS Bulletin* 24 (4): 60–72.

Deere, Carmen D., and Alain de Janvry. 1979. "A Conceptual Framework for the Empirical Analysis of Peasants." *American Journal of Agricultural Economics* 61 (4): 601–11.

Deressa, Temesgen, Rashid M. Hassan, and Claudia Ringler. 2008. "Measuring Ethiopian Farmers' Vulnerability to Climate Change across Regional States." Discussion Paper 806, Environment and Production Technology Division, International Food Policy Research Institute, Washington, DC.

Downing, Thomas. 1991. "Assessing Socioeconomic Vulnerability to Famine: Frameworks, Concepts, and Applications." Final Report to the U.S. Agency for International Development, Famine Early Warning System Project, Washington, DC.

———. 1992. "Vulnerability and Global Environmental Change in the Semiarid Tropics: Modelling Regional and Household Agricultural Impacts and Responses." Paper prepared for the International Conference on Impacts of Climatic Variations and Sustainable Development in Semi-Arid Regions, Fortaleza, State of Ceará, Brazil, January 27–February 1.

Drèze, Jean, and Amartya Sen. 1989. *Hunger and Public Action*. WIDER Studies in Development Economics. Oxford, U.K.: Clarendon Press.

Duarte, Mafalda, Rachel Nadelman, Andrew Norton, Donald Nelson, and Johanna Wolf. 2007. "Adapting to Climate Change: Understanding the Social Dimensions of Vulnerability and Resilience." *Environment Matters at the World Bank* July 2006–June 2007: 24–27.

Frank, Neil L., and S. A. Husain. 1971. "The Deadliest Tropical Cyclone in History?" *Bulletin of the American Meteorological Society* 52 (6): 438–45.

Füssel, Hans-Martin. 2007. "Vulnerability: A Generally Applicable Conceptual Framework for Climate Change Research." *Global Environmental Change* 17 (2): 155–67.

Füssel, Hans-Martin, and Richard J. T. Klein. 2006. "Climate Change Vulnerability Assessments: An Evolution of Conceptual Thinking." *Climatic Change* 75 (3): 301–29.

Gasper, Des. 1993. "Entitlements Analysis: Relating Concepts and Contexts." *Development and Change* 24 (4): 679–718.

Government of Bangladesh. 2008. "Cyclone Sidr in Bangladesh: Damage, Loss, and Needs Assessment for Disaster Recovery and Reconstruction." Dhaka, Bangladesh. http://www.preventionweb.net/files/2275_CycloneSidrinBangladesh ExecutiveSummary.pdf.

Government of Bangladesh, Ministry of Food and Disaster Management. 2008. *Super Cyclone Sidr 2007: Impacts and Strategies for Interventions.* http://www .cdmp.org.bd/reports/Draft-Sidr-Report.pdf.

Griffiths, John. 1986. "What Is Legal Pluralism?" *Journal of Legal Pluralism* 24: 1–55.

Guyer, Jane. 1981. "Household and Community in African Studies." *African Studies Review* 24 (2/3): 87–138.

Guyer, Jane, and Pauline Peters. 1987. "Introduction." Special Issue on Households. *Development and Change* 1 (18): 197–214.

Hart, Gillian. 1992. "Household Production Reconsidered: Gender, Labor Conflict, and Technological Change in Malaysia's Muda Region." *World Development* 20 (6): 809–23.

ICHRP (International Council on Human Rights Policy). 2008. *Climate Change and Human Rights: A Rough Guide.* Geneva, Switzerland: ICHRP.

Kasperson, Roger E., Kirstin Dow, Emma R. M. Archer, Daniel Cáceres, Thomas E. Downing, Tomas Elmqvist, Siri Eriksen, Carle Folke, Guoyi Han, Kavita Iyengar, Coleen Vogel, Kerrie Ann Wilson, and Gina Ziervogel. 2005. "Vulnerable Peoples and Places." In *Ecosystems and Human Well-Being, Volume 1—Current State and Trends: Findings of the Condition and Trends Working Group,* ed. Rashid M. Hassan, Robert Scholes, and Neville Ash, 143–64. Washington, DC: Island Press.

Larson, Anne M., and Jesse C. Ribot. 2007. "The Poverty of Forestry Policy: Double Standards on an Uneven Playing Field." *Sustainability Science* 2 (2): 189–204.

Leach, Melissa, and Robin Mearns. 1991. "Poverty and Environment in Developing Countries: An Overview Study." Report to the U.K. Economic and Social Research Council (Society and Politics Group and Global Environmental Change

Initiative Programme) and Overseas Development Administration. Institute of Development Studies, University of Sussex, Brighton, U.K.

Leach, Melissa, Robin Mearns, and Ian Scoones, eds. 1997. "Community-Based Sustainable Development: Consensus or Conflict?" *IDS Bulletin* 28 (4): 1–95.

———. 1999. "Environmental Entitlements: Dynamics and Institutions in Community-Based Natural Resource Management." *World Development* 27 (2): 225–47.

Leftwich, Adrian. 1994. "Governance, the State and the Politics of Development." *Development and Change* 25 (2): 363–86.

Mansuri, Ghazala, and Vijayendra Rao. 2004. "Community-Based and -Driven Development: A Critical Review." *The World Bank Research Observer* 19 (1): 1–39.

McGray, Heather, Anne Hammill, and Rob Bradley, with E. Lisa Schipper and Jo-Ellen Parry. 2007. *Weathering the Storm: Options for Framing Adaptation and Development.* Washington, DC: World Resources Institute. http://pdf.wri.org/weathering_the_storm.pdf.

Mortimore, Michael J., and William M. Adams. 2001. "Farmer Adaptation, Change and 'Crisis' in the Sahel." *Global Environmental Change* 11 (1): 49–57.

Moser, Caroline, and Andrew Norton. 2001. *To Claim Our Rights: Livelihood Security, Human Rights and Sustainable Development.* London: Overseas Development Institute.

O'Brien, Karen, Siri Eriksen, Lynn P. Nygaard, and Ane Schjolden. 2007. "Why Different Interpretations of Vulnerability Matter in Climate Change Discourses." *Climate Policy* 7 (1): 73–88.

Parry, Martin, Osvaldo F. Canziani, Jean Palutikof, Paul van der Linden, and Clair Hanson, eds. 2007. *Climate Change 2007: Impacts, Adaptation and Vulnerability. Contribution of Working Group II to the Fourth Assessment Report of the Intergovernmental Panel on Climate Change.* Cambridge, U.K.: Cambridge University Press.

Peluso, Nancy Lee. 1992. *Rich Forests, Poor People: Resource Control and Resistance in Java.* Berkeley, CA: University of California Press.

Prowse, Martin. 2003. "Toward a Clearer Understanding of 'Vulnerability' in Relation to Chronic Poverty." Working Paper 24, Chronic Poverty Research Centre, University of Manchester, Manchester, U.K.

Ramanathan, Veerabhadran, Christopher O. Justice, and Maria Carmen Lemos. 2007. *Evaluating Progress of the U.S. Climate Change Science Program: Methods and Preliminary Results.* Washington, DC: National Academies Press.

Ribot, Jesse. 1995. "The Causal Structure of Vulnerability: Its Application to Climate Impact Analysis." *GeoJournal* 35 (2): 119–22.

———. 2004. *Waiting for Democracy: The Politics of Choice in Natural Resource Decentralization.* Washington, DC: World Resources Institute.

Ribot, Jesse, Antonio R. Magalhães, and Stahis S. Panagides, eds. 1996. *Climate Change, Climate Variability, and Social Vulnerability in the Semi-Arid Tropics.* Cambridge, U.K.: Cambridge University Press.

Ribot, Jesse, and Nancy Lee Peluso. 2003. "A Theory of Access: Putting Property and Tenure in Place." *Rural Sociology* 68 (2): 153–81.

Schroeder, Richard. 1992. "Shady Practice: Gendered Tenure in The Gambia's Garden/Orchards." Paper prepared for the 88th Annual Meeting of the Association of American Geographers, San Diego, CA, April 18–20.

Scott, James C. 1976. *The Moral Economy of the Peasant: Rebellion and Subsistence in Southeast Asia.* New Haven, CT: Yale University Press.

Sen, Amartya. 1981. *Poverty and Famines: An Essay on Entitlement and Deprivation.* Oxford, U.K.: Oxford University Press.

———. 1984. "Rights and Capabilities." In *Resources, Values, and Development,* ed. Amartya Sen. Oxford, U.K.: Basil Blackwell.

Smucker, Thomas, and Ben Wisner. 2008. "Changing Household Responses to Drought in Tharaka, Kenya: Vulnerability, Persistence and Challenge." *Disasters* 32 (2): 190–215.

Solomon, Susan, Dahe Qin, Martin Manning, Melinda Marquis, Kristen Averyt, Melinda M. B. Tignor, Henry LeRoy Miller Jr., and Zhenlin Chen, eds. 2007. *Climate Change 2007: The Physical Science Basis. Contribution of Working Group I to the Fourth Assessment Report of the Intergovernmental Panel on Climate Change.* Cambridge, U.K.: Cambridge University Press.

Stern, Nicholas. 2006. *The Economics of Climate Change: The Stern Review.* Cambridge, U.K.: Cambridge University Press. http://www.hm-treasury.gov.uk/stern_review_report.htm.

Swift, Jeremy. 1989. "Why Are Rural People Vulnerable to Famine?" *IDS Bulletin* 20 (2): 8–15.

Turner, Billie Lee II, Pamela A. Matson, James J. McCarthy, Robert W. Corell, Lindsey Christensen, Noelle Eckley, Grete Hovelsrud-Broda, Jeanne X. Kasperson, Roger E. Kasperson, Amy Luers, Marybeth L. Martello, Svein Mathiesen, Rosamond Naylor, Colin Polsky, Alexander Pulsipher, Andrew Schiller, Henrik Selin, and Nicholas Tyler. 2003. "Illustrating the Coupled Human-Environment System for Vulnerability Analysis: Three Case Studies." *Proceedings of the National Academy of Sciences* 100 (14): 8080–85.

von Benda-Beckmann, Keebet. 1981. "Forum Shopping and Shopping Forums: Dispute Processing in a Minangkabau Village in West Sumatra." *Journal of Legal Pluralism* 19: 117–59.

Watts, Michael. 1983. "On the Poverty of Theory: Natural Hazards Research in Context." In *Interpretations of Calamity: From the Viewpoint of Human Ecology,* ed. Kenneth Hewitt, 231–62. London: Allen Unwin.

————. 1987. "Drought, Environment and Food Security: Some Reflections on Peasants, Pastoralists and Commoditization in Dryland West Africa." In *Drought and Hunger in Africa: Denying Famine a Future,* ed. Michael H. Glantz. Cambridge, U.K.: Cambridge University Press.

Watts, Michael, and Hans-Georg Bohle. 1993. "The Space of Vulnerability: The Causal Structure of Hunger and Famine." *Progress in Human Geography* 17 (1): 43–67.

White House. 2006. "The Federal Response to Hurricane Katrina." Washington, DC. http://georgewbush-whitehouse.archives.gov/reports/katrina-lessons-learned/.

Wisner, Ben. 1976. "Man-Made Famine in Eastern Kenya: The Interrelationship of Environment and Development." Discussion Paper 96, Institute of Development Studies, University of Sussex, Brighton, U.K.

World Bank. 1992. *Governance and Development.* Washington, DC: World Bank.

Yohe, Gary, and Richard S. J. Tol. 2002. "Indicators for Social and Economic Coping Capacity—Moving Toward a Working Definition of Adaptive Capacity." *Global Environmental Change* 12 (1): 25–40.

CHAPTER 3

Implications of Climate Change for Armed Conflict

Halvard Buhaug, Nils Petter Gleditsch, and Ole Magnus Theisen

Many high-profile individuals, nongovernmental organizations (NGOs), and policy reports have put forward alarmist claims about the enormous impacts that environmental change in general, and climate change in particular, will have on humanity. For example, a report from Christian Aid (2007) claims that an estimated 1 billion migrants between now and 2050 might "de-stabilise whole regions where increasingly desperate populations compete for dwindling food and water" (p. 2); and Homer-Dixon (2007) argues that "climate change will help produce [...] insurgencies, genocide, guerrilla attacks, gang warfare, and global terrorism." More dramatic still, a Pentagon report sketches scenarios of epic proportions, including the risk of reverting to a Hobbesian state of nature whereby humanity would be engaged in "constant battles for diminishing resources" (Schwartz and Randall 2003, p. 16). The *Stern Review* (Stern 2006) and the fourth assessment report of the United Nations Intergovernmental Panel on Climate Change (Parry et al. 2007) are much more cautious in their references to conflict, but warn against potentially dire societal consequences of climate change.

In stark contrast to such assessments, the empirical foundation for a general relationship between resource scarcity and armed conflict is indicative, at best; and numerous questions are unanswered regarding the proposed causal association between climate change and conflict. Although we cannot rule out the possibility that there is no general link between the two, major limitations in data and research designs make such a conclusion premature. In addition, the many processes associated with global warming, which truly have started to appear only over the past 15 years, have occurred during a time when we have witnessed a dramatic reduction in the frequency and severity of armed conflict. In this chapter, we first discuss trends in global climate change as well as in armed conflict. Then we look at arguments put forward as links between resource scarcity or climate change and armed conflict. Factors such as political and economic instability, inequality, poverty, social fragmentation, migration, and inappropriate responses are discussed. Third, we review and criticize the empirical literature. We conclude that our current knowledge is too limited for specific policy recommendations, and that considerable improvement is needed in both theory and testing to strengthen our knowledge of the field.

Environmental Change

Global warming is expected to bring about a number of significant changes to the environment. Among the many projected impacts highlighted in the fourth assessment report (Parry et al. 2007), we identify and discuss three potential natural consequences that could have substantial security implications.[1] The first of these is increasing scarcities of renewable resources, most notably freshwater and fertile soil. Resource scarcity is understood here as low per capita availability of a resource. Growing scarcity is generally seen as a consequence of either a dwindling resource base or an increased demand for the resource through increased population pressure or increased consumption.[2] Increasing scarcity is generally regarded as more harmful than a high level of scarcity per se. Thus, increasing resource variability, which is associated with higher levels of unpredictability, constitutes the greatest challenge to human livelihoods.

According to the fourth assessment report (Parry et al. 2007), the environmental impacts of climate change will vary enormously between regions. Some areas, such as Northern Europe, are likely to benefit from an increase in average temperature. Most parts of the world, however, face a grimmer future. Increasing temperatures, changing precipitation patterns, and an overall reduction in annual rainfall suggest that some of the most crucial subsistence resources will become increasingly scarce. This is likely to exacerbate overconsumption of groundwater in many areas, leading to depletion or possible contamination of aquifers and further reducing the supply of freshwater. A warmer climate also may result in the melting of glaciers in the Himalayas, the Andes, and several other major sources of water in the dry season for large sections of the developing world. More extreme precipitation also could increase topsoil erosion, in turn leading to less fertile soil for productive purposes and potentially turning more land into desert.

The second consequence of global warming, a rising sea level, is projected to have negative implications for peace and security through its potential for massive population displacement. Working Group II of the fourth assessment report predicts a global mean sea-level rise of between 0.28 meters and 0.43 meters in the course of this century, depending on the specific scenario chosen (p. 323). The world's coastal population (those residing below 100 meters elevation and less than 100 kilometers from the coast) is estimated (not very precisely) to rise from 1.2 billion in 1990 to between 1.8 billion and 5.2 billion by 2080 (Small and Nicholls 2003, p. 596). A rising sea level is most immediately threatening to populations on small island-states, although people in low-lying urban areas will also become more exposed to soil erosion, seasonal flooding, extreme weather, and other coastal hazards in coming decades. Unlike some anticipated climate-induced environmental changes, however, sea-level rise will occur relatively uniformly across the globe; and it is a gradual and predictable process.

The third physical consequence of climate change that is relevant to human security is natural disaster. Natural disasters can be categorized as either geologic or hydrometeorologic (climatic). In the 20th century, there was a dramatic increase in the number of reported climatic disasters.[3] This increase is often interpreted as a symptom of global warming, although it is unclear how much of the trend is driven by population growth, shifting settlement patterns, and better reporting over time.[4] In 2007, recorded disasters numbered 414, comprising nearly 17,000 fatalities (CRED 2008). Thirty-seven percent of those events occurred in Asia, accounting for 90 percent of all reported victims (including nonfatal cases). Because of the widespread character of climate-related disasters, they generate far higher numbers of victims than do geological events; but the latter are slightly more deadly. Floods constitute the most prevalent disaster type, followed by drought.

Possible Coping Strategies

Groups and societies facing dramatic reduction in the quality of life because of a changing climate can choose among several coping strategies. First, they may seek to adapt to the new challenges.[5] Adaptation can occur on any scale, from the individual to the international level, and may range from conservation programs and efforts at reducing consumption to pursuing alternative modes of livelihood. Second, if a society is unable to adjust to the new challenges, it may lapse into conflict, with one group trying to secure an increasing share of the diminishing resources—by force, if necessary.[6] Prophesied large-scale wars over oil or water (such as Klare 2001) are examples that belong to this category. People's third alternative is moving to more attractive locations when confronted by increasing climate variability and worsening environmental conditions.[7] In other words, they choose to flee rather than to fight.

Whether an increasingly exposed society seeks adaptation, conflict, or exit will depend on the nature of the changing environment, the vulnerability of the population, and contextual factors (see Barnett and Adger 2007). Gradual changes, such as desertification and sea-level rise, are generally suitable for a gradual response, including various forms of adaptation. Intensifying climate variability and natural hazards, in contrast, may exhibit much shorter temporal traits, ranging from mere minutes (landslides) to months (drought). Such environmental challenges will require almost immediate action. If the exposed population is unprepared or unable to adapt successfully, resource contention or rapid migration become more plausible outcomes. The inability to adapt plays a central role in the environmental security literature. Homer-Dixon (1991, 1999) labels this the "ingenuity gap"—the gap between those able and those unable to address resource scarcity by innovation. It is this gap that is promoted as an explanation for why developing countries are more prone than more-developed nations to instability and conflict stemming from scarce resources.

Armed Conflict

Figure 3.1, which is based on the Uppsala Conflict Data Program/International Peace Research Institute of Oslo Armed Conflict Dataset, shows that the number of conflicts in the world increased steadily from the mid-1950s until it peaked in the early 1990s. Since then, armed conflicts have become significantly less frequent.[8] The severity of armed conflict, measured as annual battle-related deaths, is influenced strongly by individual wars; but it has generally declined since World War II.[9]

Figure 3.1. Frequency and Severity of Armed Conflict, 1946–2007

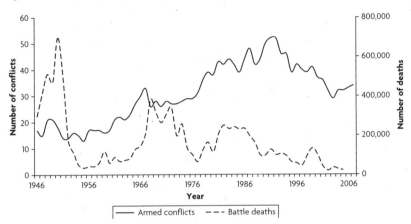

Sources: UCDP/PRIO (2008), v. 4-2008 (Gleditsch et al. 2002); PRIO Battle Deaths Dataset, v. 2.0 (Lacina and Gleditsch 2005). Data are available from http://www.prio.no/cscw/datasets and http://www.pcr.uu.se/research/UCDP/.

Although the annual incidence of conflict dropped substantially during the first decade after the end of the Cold War, the trend now appears to have leveled. This leveling is not the result of an increase in new conflicts, but of a resurgence of old ones and a decline in the rate of successful conflict resolution. More than half of today's conflicts originated during the Cold War, and the average conflict is becoming older.

Figure 3.2 shows countries in armed conflict in 2007 (dark gray) and countries with intrastate armed conflict on their soil between 1989 and 2006 (light gray). The symbols are placed at the approximate subnational center of conflict zones. The map also reveals that almost all of today's conflict-ridden countries have at least one neighboring country in conflict.

Linking Climate Change to Armed Conflict

One caveat is warranted before discussing a connection between conflict and climate trends: A comparison between the temporal patterns in armed conflict and in global warming (figure 3.3) reveals radically opposing trends since 1990. In statistical terms, the post-Cold War correlation

Figure 3.2. Intrastate Armed Conflicts, 1989–2007

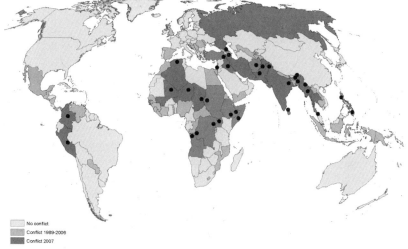

Source: UCDP/PRIO 2007.

Figure 3.3. Trends in Global Warming and Armed Conflict, 1946–2007

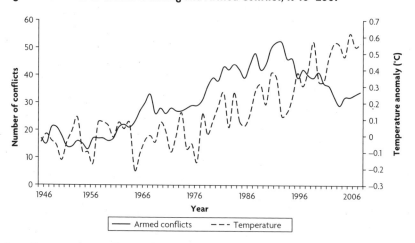

Source: Temperature statistics (deviation from global mean 1951–80) from the National Aeronautics and Space Administration Goddard Institute for Space Studies, Columbia University, New York (http://data.giss.nasa.gov/gistemp).

between conflict and temperature deviation is negative and statistically significant. Such a bivariate assessment should be interpreted with caution, but it serves to call for similar caution when claiming a causal connection between climate variability and armed conflict.

Climate Change and Conflict—A Synthesized Causal Model

In contrast to simplistic media portrayals, few scholars claim a direct link between resource scarcity and armed conflict.[10] Rather, most assessments of the environment and conflict sketch a causal story where scarcity of renewable resources, or a severe natural disaster, adds yet another stone to the burden. Thus, violence is a probable outcome only in societies already suffering from a multitude of other ills. For instance, Homer-Dixon (1999) writes, "environmental scarcity is *never* [emphasis added] a sole or sufficient cause of large migrations, poverty, or violence; it always joins with other economic, political, and social factors to produce its effects" (p. 16).

At least five social effects of climate change have been suggested as crucial intermediating catalysts of organized violence. First, reduced state income resulting from increased resource scarcity may hinder the delivery of public goods, reduce political legitimacy, and give rise to political challengers (Homer-Dixon 1999). Second, increasing resource competition in heterogeneous societies may attract opportunistic elites who intensify social cleavages—particularly ethnic identities—and make the population more vulnerable to being radicalized (Kahl 2006). Third, increasing scarcity of renewable resources in subsistence-economy societies may cause unemployment, loss of livelihood, and loss of economic activity (Ohlsson 2003), thus also decreasing state income (Homer-Dixon 1999). Fourth, efforts to adjust to a changing climate—or to remove the causes of global warming—may have inadvertent side effects that could stimulate tension and conflict. Giving the climate change issue a high profile also may serve as a scapegoat or rallying point for actors with hidden agendas (Salehyan 2008; Goldstone 2001). Finally, deteriorating environmental conditions may force people to migrate in large numbers, thereby increasing environmental stress in the receiving area and raising the potential for radicalization and ethnic hatreds (Reuveny 2007). Whether adverse climatic changes result in any of these social effects depends largely on the characteristics of the affected area. Economically developed and politically stable societies are well able to handle and adapt to conceivable environmental conditions. We should not expect an influx of environmental migrants, increasing climate variability, and sea-level rise to constitute a significant security threat in these countries. In contrast, countries that are characterized by other conflict-promoting features—notably, poor governance, large populations with polarized subgroups, social inequalities, a violent neighborhood, and a history of violence—are plausible candidates for climate-induced conflict. Figure 3.4 gives a visual impression of the synthesized causal model, and the following sections expand on each of the five suggested mechanisms.

Figure 3.4. Possible Pathways from Climate Change to Conflict

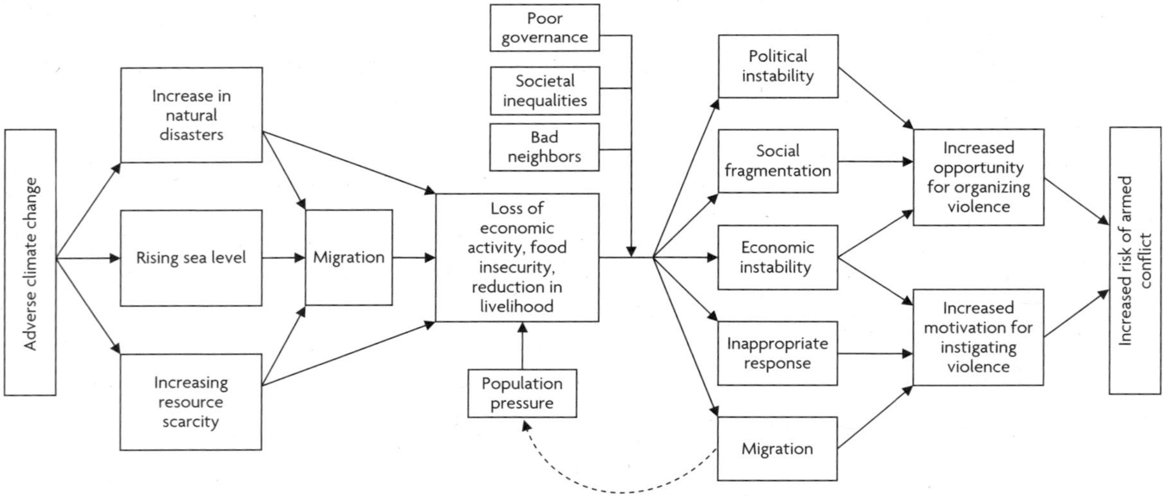

Source: Authors' illustration.

The figure gives a synthesized account of proposed causal links between climate change and armed conflict. For the sake of clarity, possible feedback loops, reciprocal effects, and contextual determinants are kept at a minimum.

Political Instability. Many scarcity-based accounts of armed conflict point to the weakening of the state as an important intermediate development. First, responding to soil degradation, crop failure, or drought is costly; and the poorest and institutionally weakest regimes simply may not be able respond in a manner that satisfies the disgruntled population. Second, increasing climatic variations may affect the redistributive capacity of governments and drain attention and capital away from other important social programs, including health, education, infrastructure, and security. Regimes also may seek to gain politically from adverse environmental developments by playing social groups against each other. Such "state exploitation" behavior has been argued to be a characteristic of several resource conflicts, including election-related violence in Kenya in 1992 and the genocide in Rwanda in 1994 (Kahl 2006). A weakened state also may give rise to opportunistic challengers who do not suffer from worsened environmental conditions. Finally, political elections in systems with little tradition of democratic rules of government are associated with higher levels of uncertainty and a higher risk of violence (Strand 2007).

There is substantial empirical evidence for a connection between political instability and increased risk of armed conflict. In a comprehensive empirical evaluation of a very wide range of proposed conflict-inducing factors in the quantitative literature, Hegre and Sambanis (2006) find recent political instability to be among the relatively few robust correlates of civil war. A number of studies on the relationship between civil war and level of democracy find support for a curvilinear relationship (for example, Hegre et al. 2001 and Muller and Weede 1990), at least in part related to the greater instability of regimes between full democracy and full autocracy (Gleditsch, Hegre, and Strand 2009; Vreeland 2008). Democratic systems avoid public unrest primarily through justice, responsiveness, and protection of minority rights; autocratic states deter organized violence primarily by denying the formation of effective opposition groups and by widespread repression. Another explanation for this curvilinear relationship is that rulers of in-between regimes are too weak to gain the control needed to become autocrats, leading to some democratic concessions (Fearon and Laitin 2003).

Thus, both ideal-type regimes are less vulnerable to climate change than are inconsistent political systems, everything else being equal.

Social Fragmentation. The connection between ethnicity and armed conflict remains a hotly debated topic in academia. Some claim that heterogeneity increases the baseline risk of civil war (Blimes 2006); others argue that particular configurations of the heterogeneous population are important (for example, dominance [Collier and Hoeffler 2004]; polarization [Esteban and Ray 2008; Reynal-Querol 2002]; and exclusion [Buhaug, Cederman, and Rød 2008; Cederman and Girardin 2007]); some argue that ethnicity is irrelevant to the onset of conflict (Fearon, Kasara, and Laitin 2007); and some reject the very notion of a given static ethnic identity (Bowen 1996). So far, systematic, statistical research has not succeeded in converging on this issue.

Regardless of the origin of civil war, ethnicity is widely regarded as a facilitator for mobilization; and language, religion, and nationality often serve as lines of demarcation between contending groups. There is often a mutually reinforcing relationship between ethnic identity and hostilities. For example, Gurr (2000) argues that the notion of a shared Eritrean nationality grew only slowly during the prolonged war of independence—being a product of the conflict rather than a precondition for it. Similarly, Prunier (2007) and Suliman (1997) emphasize the historically low ethnic barriers in Darfur, which arguably have been heightened by the prolonged conflict there. This underscores the importance and feasibility of taking effective action at an early stage in local resource conflicts, before cultural differences become "tectonic" fault lines.

Climate change is not likely to affect the ethnic composition of countries in the short run, although it may be necessary to make an exception for rapid, if temporary, disaster-induced population displacement. In a slightly longer-term perspective, however, we may witness substantial intra- and interstate migration as global warming makes environmentally vulnerable areas less sustainable.[11]

Poverty, Inequality, and Economic Instability. A second climate-induced catalyst of social instability and armed conflict is economic instability and stagnation. Food insecurity and loss of livelihood are possible consequences of adverse climatic changes in many parts of the world, resulting in poverty at the national and individual levels. Poverty (typically defined as low per capita income) long has been considered a major cause of civil war (Collier et al. 2003), and its general relative effect and statistical robustness are paralleled only by population size (Hegre and Sambanis 2006). There are several partly overlapping explanations for this fact.

Political economists usually attribute the poverty-conflict association to factors that increase individuals' inclination to criminal behavior, relative to normal economic activity (for example, Collier and Hoeffler 2004; Berdal and Malone 2000; and Grossmann 1991). Put differently, poor opportunities for legal income earning (that is, low wages, high unemployment rate) lower the threshold for joining a rebellion. In addition, loss of income may force affected people to migrate—and that constitutes a separate, indirect potential for population pressure, resource competition, and rebel recruitment (Raleigh, Jordan, and Salehyan 2008; chapter 4 of this volume).

A more traditional explanation in the conflict literature stresses the motivational aspect of rebellion. Here, civil conflict is understood as a product of relative deprivation, where increasingly marginalized segments of society take up arms to alter the status quo (Gurr 1970). Social inequality can be categorized as either vertical or horizontal. Vertical inequality denotes systematic differences in opportunities and privileges between the worst-off and best-off individuals. Most statistical research on inequality focuses on vertical differences (Collier and Hoeffler 2004; Hegre, Gissinger, and Gleditsch 2003). In contrast, horizontal inequality taps systematic differences in opportunities and privileges between different groups. Although both forms of inequality may give rise to social unrest, intergroup differences now are regarded as more conflict prone than inequalities between social classes (Østby 2006, 2008; Stewart, Brown, and Mancini 2005; Stewart 2000). Empirical support for the horizontal inequality-conflict thesis includes Besançon (2005); Murshed and Gates (2005); and Østby, Nordås, and Rød (2009).

Supplementing the economic individualist and sociological group-level explanations of the poverty-conflict nexus, political scientists often maintain that a low per capita income is a symptom of weak state capacity (for example, Fearon and Laitin 2003). According to this perspective, the concentration of civil wars in poor countries is explained as much by the favorable conditions for insurgency (that is, poor counterinsurgency capability, limited infrastructure, and lack of local governance) as by a rational calculus of opportunistic individuals. Furthermore, poor economic performance erodes popular support for the regime. In consolidated democracies and harsh authoritarian systems, this may not be harmful; but in unstable and democratizing states, a decline in political legitimacy easily translates into a distrust of the political system at large and provides opportunities for nondemocratic challengers.

A rapidly changing climate will have the largest short-term effects on economies dependent on production/exports of renewable primary commodities. Some societies will be better off, but many countries in the developing world will experience reduced agricultural yields. This, in turn, may contribute to an increasing ingenuity gap between developed and developing countries, whereby the latter have less to spend on such adaptive buffers as resilient infrastructure, irrigation systems, and desalinization plants for freshwater generation (Homer-Dixon 1999). To the extent that environmental changes will vary substantially within countries, they also are likely to amplify existing intergroup inequalities.

Migration. Migration may be both a cause and an effect of worsening environmental conditions; hence, it enters the causal model (figure 3.4) at two stages. Migration can be *rapid* or *gradual*, reflecting the speed of the emerging perceived environmental push and pull factors. It can be *permanent* or *temporary*, also reflecting the nature of the threat. (Sudden alterations, such as those from natural disasters, are more likely to cause temporary displacement.) Migrants may also be separated into those who move only as far as necessary to avoid the immediate danger and those who travel long distances and (attempt to) settle in the safe, developed world. Furthermore, the underlying mechanisms of refugee flight (when the risk to one's life makes leaving imperative) and conventional migration (when staying is a viable option) are qualitatively different. For example, inhabitants of small island-states ultimately have to relocate if some of the more dire predictions about future sea-level rise come true, whereas people living in increasingly dry areas may have less extreme adaptation strategies at hand. Moreover, there may be numerous overlapping environmental, political, and economic push factors, as well as pull factors in the receiving area, that influence the decision whether, where, and when to move.

Climate-induced migration is argued to lead to violent conflict in receiving areas through at least four complementary processes (Reuveny 2007). First, the arrival of newcomers can lead to competition over diminishing natural and economic resources, especially if property rights are underdeveloped. Second, a wave of migrants of a different ethnic origin than the local population may give rise to ethnic tension and solidification of identities. Third, large flows of migrants may cause mistrust between the sending and receiving states. Finally, climate-induced migration may create or exacerbate traditional fault lines—for instance, when migrant pastoralists and local sedentary farmers compete over the use of land. Although there is some evidence for a link between transnational refugee flows and the outbreak of armed conflict (Buhaug and Gleditsch 2008; Salehyan 2007), it is not obvious that environment-induced population flows will have the same security implications for the host population as do migrants escaping armed violence. Because of data limitations and lack of conceptual clarity, no empirical study has been able to explore the general consequences of "environmental migration" across cases.

This area remains critically understudied, and it is not clear whether we should expect climate-induced migration to blend with the ubiquitous urbanization or follow a radically different path.

Inappropriate Response. A final but less acknowledged potential catalyst of social friction and armed violence concerns inadvertent consequences of human reactions to climate change. At a macro level, draconian measures to reduce carbon dioxide emissions may have large, unforeseen (or underestimated) effects on global and regional economic systems. For instance, in high-growth developing countries like China and India, such measures might make trading-state strategies less attractive than strategies of territorial warfare (see Rosecrance 1986). Also, enforcing drastic emissions-cutting regulations very likely would lead to a stagnation or even recession of such economies, with political instability and civil unrest as probable outcomes. Heavy taxation on international air travel could have negative impacts on small, tourism-dependent economies.

On a smaller scale, dam building and development of irrigation systems to counter projected changes in precipitation patterns may have inadvertent consequences, especially for downstream populations. The construction of such installations also may have direct adverse environmental effects on local communities; and if these effects are not properly compensated, there is a significant potential for protest and conflict (see Baechler 1999). Moreover, the expansion of biofuel programs could have serious implications for the regional (if not necessarily global) food situation. In the last few years, in fact, there has been a sharp rise in food prices after decades of decline (Gleditsch and Theisen 2009; IFPRI 2007). In Mexico, a reported 70,000 people took to the streets to protest mounting tortilla prices caused by increasing U.S. demands for Mexican corn for biofuel production (Watts 2007). So far, research on the security implications of climate change has not paid much attention to these potential catalysts.

Additionally, climate change could influence how armed conflicts are perceived and justified. In illiberal regimes, global warming may constitute a much-needed political escape because no single country is to blame for the adverse environmental developments and no country can be expected to mitigate these problems single-handedly. A relevant example is provided by the current debate on causes of the Darfur conflict. Some, including the regime in Khartoum, claim that the origins of the conflict can be traced to the decades-old Sahelian drought (see, for example, UNEP 2007). Such claims may have some merit (however, see de Waal 2007), but they are highly problematic because they suggest a near-deterministic relationship between the environment and armed conflict—thereby relieving the main actors of their own responsibility. In fact, even the United Nations has been accused of using climate change as an excuse for its inability to halt the killings in Darfur (Crilly 2007). Moreover, the high profile of the climate issue entails a high risk for political actors with hidden agendas who rally around the popular notion of global warming constituting the greatest security challenge of our time (see Salehyan 2008 and Goldstone 2001).

Contextual Mediators

The potential mechanisms whereby climate change and variability increase the risk of conflict, as laid out above, are not likely—or even plausible—to play out in all societies. In fact, many of the same social entities that risk being affected negatively by changing environmental conditions also determine the likelihood and extent to which such consequences will materialize.

General research on civil war has shown that the risk of conflict is associated—significantly and robustly—with low national income, a large population, weak and inconsistent political institutions, an unstable neighborhood, and a recent history of violence.[12] Negative security impacts of future climatic changes are likely to be observed primarily in countries and regions that host today's armed conflict: the east-central parts of Africa, the Middle East, and Central and East Asia. Accordingly, it is those areas where international peace-building and development efforts should have their centers of gravity.

The physical consequences of climate change, like many contemporary armed conflicts, do not follow state boundaries. There are several complementary explanations for the clustering of armed conflict, including proliferation of small arms and know-how, transnational ethnic links, refugee flows, and negative economic externalities of violence (Buhaug and Gleditsch 2008). However, whereas the extent and intensity of civil war are influenced by such factors as type of conflict and fighting capacity, population distribution and relocation, cultural delineation, terrain, and the location of strategically important features, climatic variations tend to follow topographic and meteorologic boundaries and macro-level climatic patterns. Just as the regional context may affect whether adaptation turns violent, adverse local changes also may have negative spillovers throughout the region. Accordingly, concerted efforts at addressing potential, harmful societal consequences of climate change also need to apply a regional perspective. It makes little sense to invest heavily in peace making and development in Chad without simultaneously handling the situation in neighboring Sudan. And the Kurdish question in Iraq cannot be discussed without acknowledging the role of neighboring Turkey and the Islamic Republic of Iran. Isolated peace-building initiatives in unstable areas of the world are suboptimal, at best. More likely, they are unsuccessful. Hence, understanding the regional dynamics of conflict and human security is vital in securing sustainable development and solidification of the political system in societies emerging from armed conflict.

Assessing the Empirical Literature

In contrast to the rich causal stories presented in the case literature, statistical comparative studies on the subject tend to model the scarcity-conflict relationship in rather simplistic ways. The first true multivariate assessment, conducted by Hauge and Ellingsen (1998), found that land degradation, freshwater scarcity, population density, and deforestation all have direct, positive effects on the incidence of civil conflict. Although no interactive effects were reported, the authors mention in a footnote that no such indirect associations were uncovered. In contrast, the contemporaneous Phase II of the State Failure Task Force (SFTF; now the Political Instability Task Force) project (Esty et al. 1998) did not find any direct relationship between indicators of environmental scarcity and state failure. Although differences in data and research designs provide some explanation for the lack of correspondence between these pioneering studies, Theisen (2008) shows that Hauge and Ellingsen's results cannot be reproduced, even with the original data. Later attempts that focus on water availability are somewhat more coherent. Miguel, Satyanath, and Sergenti (2004) find that negative deviation in annual precipitation in sub-Saharan Africa substantially reduces national economic growth in these countries, and thereby indirectly increases the risk of civil war. Similar findings are reported by Hendrix and Glaser (2007) and Levy et al. (2006). Raleigh and Urdal (2007) also find water scarcity and soil degradation—two likely outcomes of climate change—to increase the risk of armed conflict; but the general effects of resource scarcity are less pronounced in developed countries than in developing ones, contrary to virtually all theoretical arguments on the topic.

Other studies have focused more explicitly on how population pressure (high population density, high population growth, and large youth cohorts) relates to civil conflict. For example, Urdal (2005) finds that high pressure on potential cropland is related *negatively* to civil conflict; but that population growth and density jointly increase the risk of conflict, if only marginally. De Soysa (2002a,b) and Raleigh and Urdal (2007) report a positive effect of population density on the baseline risk of conflict. Others—notably, Hegre and Sambanis (2006) in their comprehensive statistical analysis—provide no support for the neo-Malthusian population-pressure hypothesis. Some studies even find empirical evidence directly opposing the causal pathways proposed in the case-based environmental security literature. For example, de Soysa (2002b) finds that rural population density combined with renewable resource *wealth* increase the conflict risk. Similarly, Binningsbø, de Soysa, and Gleditsch (2007) report that higher levels of accumulated consumption of renewable resources (the so-called ecological footprint) are associated with a lowered risk of civil conflict, even after controlling for economic development. And in the most comprehensive reevaluation of the topic to date, Theisen (2008) demonstrates that several earlier findings are either not replicable or do not hold with improved data.

Research on natural disasters and conflict is less developed, and only a handful of recent studies have used quantitative models on a large number of cases. So far, this literature is quite supportive of the general disaster-conflict hypothesis (for example, Nel and Righarts 2008, Brancati 2007, and Drury and Olson 1998), although results appear to be driven primarily by geological rather than climatic events. Moreover, little is known about the mechanisms that explain this correlational pattern—whether it arises from frustration stirred by inadequate governmental assistance, loss of public goods delivery, and a failing national economy or is a result of opportunistic actors taking advantage of the state's weakened counterinsurgency capability.

Despite a lack of convergence on any of the important proposed links, it would be premature at this time to conclude that environmental factors generally are unrelated to conflict risk. Underdeveloped theoretical models, poor data quality, and inappropriate research designs cast doubt on possible inferences to be drawn from the reported (non)findings. We identify five crucial limitations in the current literature that should be addressed in future work in this area.

The first problem is that large-N studies have largely failed to account adequately for proposed indirect and conditional effects of climate change. Of the mediating factors presented in figure 3.4, only simple interaction terms have been used, at best. However, there is no a priori reason why such intermediate effects should be linear and multiplicative in nature; rather, they may be characterized by threshold effects and only apply under certain conditions. Furthermore, the interactive effects might consist of a complex web of multiple factors, thus making it harder to test environmental security arguments with conventional statistical methods in the absence of a well-developed theoretical model. A natural first step would be to isolate the analysis to settings where organized violence is a plausible outcome under increasing environmental stress. A complementary option would be to use the advantages of so-called fuzzy methods, which are ideal when dealing with multiple interactive factors that together bring an outcome (Ragin 1987).

Second, large-N investigations of armed conflict suffer from overly aggregated research designs (O'Lear and Diehl 2007; Buhaug and Lujala 2005). This is even more the case for research on environmental scarcity and armed conflict. In contrast, unrest rarely engulfs entire states, and many conflict-ridden countries today appear relatively unaffected by the violence (for example, India, Thailand, and Turkey). By studying the environment-conflict relationship at the country level, one encounters a significant risk of creating ecological fallacies—or explaining local phenomena from aggregated data. Recent advances in geographic information systems and remote sensing imply that disaggregated statistical analyses now are becoming a viable option (Raleigh and Urdal 2007).

A third significant limitation concerns the dependent variable. Almost all statistical assessments of environmental scarcity and conflict look only at the most severe forms of organized violence: civil and interstate war. Arguably, however, these conflicts are the least likely to emerge from an increasing scarcity of renewable resources. The cost of fighting a government army is considerable, so deprived groups simply may be too weak to engage in serious conflict with state forces (Klare 2001). This suggests that violent conflicts between groups, without direct involvement of the state, will be the most frequent form of conflict under worsening climatic conditions. Indeed, much of the case literature that claims a causal link between scarcities of renewable resources and armed violence refer to local, small-scale interethnic conflicts (Kahl 2006; Martin 2005; Suliman 1999). Future research also should pay attention to the influence of the environment on conflict dynamics (duration, severity, and diffusion)—until now, the topic has been overlooked almost completely.

Fourth, research in the field has long suffered from a dearth of reliable environmental data. For example, the SFTF phase II report (Esty et al. 1998) admits that data on water quality were available for only 38 countries, and indicators of other forms of scarcity varied greatly in coverage. In fact, the SFTF's conclusions regarding the role of environmental factors were based on statistical models that exclude nearly half of the world's independent states because of missing data. The temporal dimension—so crucial if we are to gauge the dynamics of resource availability—is another inherent challenge. In Hauge and Ellingsen's (1998) pioneering study, measures of both soil degradation and freshwater availability are entirely static, severely restricting the breadth of possible inferences. The climate debate, in contrast, is all about *changes* to current environmental conditions. A third data limitation concerns the level of spatial resolution. Environmental data are often aggregated and released at the country level; but the underlying data may well be collected at a higher resolution. In other cases, researchers fail to exploit the richness of available subnational data. In their widely cited study, Miguel, Satyanath, and Sergenti (2004) construct a country-level proxy for economic shock from geo-referenced rainfall data that were collected at a much higher degree of spatial resolution.

Finally, the case study literature tends to select cases on the dependent and main independent variables; that is, it tends to study only countries where both conflict and scarcity are prevalent. Gleditsch (1998) also argues that much of this literature suffers from complex and untestable models, confusion of level of analysis, and inference based on speculations and anecdotal evidence. Other criticisms point to a possible spurious link between resource scarcity and armed conflict, according to which both are caused by some third factor. The lack of focus on the agency or stated objective of the actors involved in the conflict also has been criticized (Peluso and Watts 2001). In addition, the sometimes explicit refusal to attempt to rank the causal factors (for example, in Schwartz, Deligiannis, and Homer-Dixon 2001 and Homer-Dixon 1999) puts significant limitations on inference, as well as effective policy recommendations.

Conclusion

Climate change has many probable consequences for our physical environment, and each of these has a variety of potential consequences for human livelihoods. The scope of these challenges to human societies adds to the urgency of the climate change debate. There is hardly a single facet of our daily lives that does not have some latent effect on our ability to deal with climate change. The good news is that support for policies of mitigation and adaptation can mobilize a broad section of the public because climate change is relevant virtually to everybody. The bad news is that such broad debates are in danger of being hijacked by people with special agendas. We believe that this has happened, to some extent, in the debate about climate change and conflict, where NGOs and security establishments have vested interests in presenting their own mission as one that also is particularly well suited to the climate change agenda. Accordingly, some level of caution is warranted.

In great contrast to popular conceptions—which have been boosted by massive media attention, alarmist claims by high-profile actors, and the 2007 Nobel Peace Prize award—the empirical foundation for a general relationship between resource scarcity and armed conflict is indicative, at best; and numerous questions remain to be answered regarding the exact nature of the proposed causal association between climate change and conflict. Case-based research offers several examples of conflicts where environmental degradation (both that caused by humans and that resulting from climate variability) plausibly has had some influence on the initiation of violence. Although environmental problems abound, however, armed conflict is a very rare phenomenon; at present, there does not appear to be a general link between the two. In fact, the recent, accelerating global warming strongly contradicts contemporaneous developments in the global frequency and severity of armed conflict. To the extent that policy advice from the academic community should be founded on robust findings in peer-reviewed research, the literature on environmental conflict has surprisingly little to offer. We are only beginning to experience the physical changes imposed by global warming, however, so a lack of systematic association between the environment and armed conflict today need not imply that such a connection cannot materialize tomorrow.

There is considerable room for improvement within the case study tradition. In our view, however, the most important immediate challenges lie within generalizable, statistical research. Case studies can provide some advance warning of deteriorating conditions in selected areas, but global climate change policy is crucially dependent on giving early warning of events in areas that may not have had such problems in the past. For this, we need better generalizable knowledge. Data limitations, rigid research designs, and overly bold assumptions have effectively prevented direct and thorough evaluations of prevailing causal theories. These challenges are not insurmountable. Recent and ongoing advances in data collection and statistical software—notably within geographic information systems—coupled with constant refinement of theoretical models facilitate more precise and localized analysis of environment-conflict links than what has been published to date. Eventually, a multidisciplinary research program, combining the best of the two research traditions, would provide the best foundation for an assessment of the security implications of climate change.

Notes

1. This chapter applies a traditional definition of "security," meaning the absence of armed conflict. The terms "conflict" and "armed conflict" are used interchangeably.
2. Homer-Dixon (1999) uses the concept of environmental scarcity. In addition to the two components we include, the concept also incorporates distributional issues. For purposes of analytical clarity, we exclude the distributional aspect from our definition and discussion of scarcity here, but return to the issue of inequality in the discussion of how resource scarcity can translate into armed conflict.
3. The Centre for Research on the Epidemiology of Disasters defines a natural disaster as "a situation or event, which overwhelms local capacity, necessitating a request to national or international level for external assistance; an unforeseen and often sudden event that causes great damage, destruction, and human suffering" (CRED 2008, p. 2). A disaster is entered into CRED's database if at least 10 people are reported killed, 100 people are reported affected, a state of emergency has been declared, or a call for international assistance has been issued.

4. The severity of disasters, measured by the number of casualties, shows no evident time trend. That fact could be seen as evidence of improvement in coping capabilities in disaster-prone societies, although it also reflects general population growth, shifting settlement patterns, and possibly a reporting bias in that lesser disasters were less likely to be registered early in the period.

5. We understand adaptation as "adjustment in natural or human systems in response to actual or expected climatic stimuli or their effects, which moderates harm or exploits beneficial opportunities" (Parry et al. 2007, p. 6).

6. For a discussion, see Homer-Dixon (1999).

7. Mitigation efforts, or initiatives to remove the causes of anthropogenic climate change (for example, carbon capture and storage and the development of alternative energy sources) are inherently global in scope. Furthermore, the momentum of current global warming implies that such strategies are viable only from a long-term perspective. We do not discuss this issue further here.

8. The Armed Conflict Dataset project defines armed conflict as "a contested incompatibility that concerns government or territory or both where the use of armed force between two parties [of which at least one is the government of a state] results in at least 25 [annual] battle-related deaths" (Gleditsch et al. 2002, pp. 618–19). Data are available at http://www.prio.no/cscw/armedconflict/and at http://www.pcr.uu.se.

9. These findings are according to the International Peace Research Institute of Oslo's Battle Deaths Dataset, v. 2.0 (Lacina and Gleditsch 2005). Data are available at http://www.prio.no/cscw/armedconflict/. The battle-deaths data include all reported killings in battles (including civilians killed in crossfire) between the recorded actors. These figures do not include indirect casualties resulting from hunger and epidemics in the wake of conflict, nor do they include casualties from one-sided violence (genocide, ethnic cleansing, terrorism), conflicts between groups but not involving the government, and criminal behavior.

10. An exception is sometimes made for freshwater; and the Middle East and North Africa appear frequently in the literature on "water wars" (for example, Gleick, Yolles, and Hatami 1994). So far, however, no international water dispute has escalated to the level of war. Moreover, research suggests that although sharing a river basin increases the probability of low-level disputes between states (Gleditsch et al. 2006), it also increases the level of cooperation (Brochmann and Gleditsch 2006).

11. Estimates of the number of "environmental refugees" abound, though few seem to be based on transparent and sound calculations. For example, a report by Christian Aid (2007) predicts that 1 billion people will be forced from their homes before 2050, mostly because of changes resulting from economic development.

12. For a review, see Hegre and Sambanis (2006).

References

Baechler, Gunther. 1999. *Violence through Environmental Discrimination: Causes, Rwanda Arena, and Conflict Model.* Dordrecht, Netherlands: Kluwer Academic.

Barnett, Jon, and W. Neil Adger. 2007. "Climate Change, Human Security and Violent Conflict." *Political Geography* 26 (6): 639–55.

Berdal, Mats R., and David M. Malone. 2000. *Greed and Grievance: Economic Agendas in Civil Wars.* Boulder, CO: Lynne Rienner.

Besançon, Marie. 2005. "Relative Resources: Inequality in Ethnic Wars, Revolutions, and Genocides." *Journal of Peace Research* 42 (4): 393–415.

Binningsbø, Helga Malmin, Indra de Soysa, and Nils Petter Gleditsch. 2007. "Green Giant, or Straw Man? Environmental Pressure and Civil Conflict, 1961–99." *Population and Environment* 28 (6): 337–53.

Blimes, Randall J. 2006. "The Indirect Effect of Ethnic Heterogeneity on the Likelihood of Civil War Onset." *Journal of Conflict Resolution* 50 (4): 536–47.

Bowen, John R. 1996. The Myth of Global Ethnic Conflict." *Journal of Democracy* 7 (4): 3–14.

Brancati, Dawn. 2007. "Political Aftershocks: The Impact of Earthquakes on Intrastate Conflict." *Journal of Conflict Resolution* 51 (5): 715–43.

Brochmann, Marit, and Nils Petter Gleditsch. 2006. "Shared Rivers and International Cooperation." Paper presented at the Workshop on Polarization and Conflict, Nicosia, Cyprus, April 26–29.

Buhaug, Halvard, Lars-Erik Cederman, and Jan Ketil Rød. 2008. "Disaggregating Ethno-Nationalist Civil Wars: A Dyadic Test of Exclusion Theory." *International Organization* 62 (3): 531–51.

Buhaug, Halvard, and Kristian Skrede Gleditsch. 2008. "Contagion or Confusion? Why Conflicts Cluster in Space." *International Studies Quarterly* 52 (2): 215–33.

Buhaug, Halvard, and Päivi Lujala. 2005. "Accounting for Scale: Measuring Geography in Quantitative Studies of Civil War." *Political Geography* 24 (4): 399–418.

Cederman, Lars-Erik, and Luc Girardin. 2007. "Beyond Fractionalization: Mapping Ethnicity onto Nationalist Insurgencies." *American Political Science Review* 101 (1): 173–85.

Christian Aid. 2007. *Human Tide: The Real Migration Crisis. A Christian Aid Report.* London: Christian Aid.

Collier, Paul, V. Lance Elliott, Håvard Hegre, Anke Hoeffler, Marta Reynal-Querol, and Nicholas Sambanis. 2003. *Breaking the Conflict Trap: Civil War and Development Policy.* World Bank Policy Research Report. Washington, DC: World Bank.

Collier, Paul, and Anke Hoeffler. 2004. "Greed and Grievance in Civil War." *Oxford Economic Papers* 56 (4): 563–96.

CRED (Centre for Research on the Epidemiology of Disasters). 2008. "Annual Disaster Statistical Review: The Numbers and Trends 2007." Centre for Research on the Epidemiology of Disasters, Brussels, Belgium.

Crilly, Rob. 207. "Darfur Conflict Is 'Warning to World' of Climate Change Peril." *The Times*, June 23. http://www.timesonline.co.uk/tol/news/world/africa/article 1975132.ece.

de Soysa, Indra. 2002a. "Ecoviolence: Shrinking Pie or Honey Pot?" *Global Environmental Politics* 2 (4): 1–34.

———. 2002b. "Paradise Is a Bazaar? Greed, Creed, and Governance in Civil War, 1989–99." *Journal of Peace Research* 39 (4): 395–416.

de Waal, Alex. 2007. "Sudan: What Kind of State? What Kind of Crisis?" Occasional Paper 2, Crisis States Research Centre, London.

Drury, Alfred C., and Richard S. Olson. 1998. "Disasters and Political Unrest: A Quantitative Investigation." *Journal of Contingencies and Crisis Management* 6 (3): 153–61.

Esteban, Joan, and Debraj Ray. 2008. "Polarization, Fractionalization, and Conflict." *Journal of Peace Research* 45 (2): 163–82.

Esty, Daniel C., Jack A. Goldstone, Ted Robert Gurr, Barbara Harff, Marc Levy, Geoffrey D. Dabelko, Pamela T. Surko, and Alan N. Unger. 1998. "State Failure Task Force Report: Phase II Findings." Science Applications International, McLean, VA.

Fearon, James D., Kimuli Kasara, and David D. Laitin. 2007. "Ethnic Minority Rule and Civil War Onset." *American Political Science Review* 101 (1): 187–93.

Fearon, James D., and David D. Laitin. 2003. "Ethnicity, Insurgency and Civil War." *American Political Science Review* 97 (1): 75–90.

Gleditsch, Nils Petter. 1998. "Armed Conflict and the Environment. A Critique of the Literature." *Journal of Peace Research* 35 (3): 381–400.

Gleditsch, Nils Petter, Kathryn Furlong, Håvard Hegre, Bethany Lacina, and Taylor Owen. 2006. "Conflicts over Shared Rivers: Resource Scarcity or Fuzzy Boundaries?" *Political Geography* 25 (4): 361–82.

Gleditsch, Nils Petter, Håvard Hegre, and Håvard Strand. 2009. "Democracy and Civil War." In *Handbook of War Studies III: The Intrastate Dimension*, ed. Manus I. Midlarsky. Ann Arbor, MI: University of Michigan Press.

Gleditsch, Nils Petter, and Ole Magnus Theisen. 2009. "Resources, the Environment, and Conflict." In *Routledge Handbook of Security Studies*, ed. Victor Mauer and Myriam Dunn Cavelty, 220–33. London: Routledge.

Gleditsch, Nils Petter, Peter Wallensteen, Mikael Eriksson, Margareta Sollenberg, and Håvard Strand. 2002. "Armed Conflict 1946–2001: A New Dataset." *Journal of Peace Research* 39 (5): 615–37.

Gleick, Peter H., Peter Yolles, and Haleh Hatami. 1994. "Water, War, and Peace in the Middle East." *Environment* 36 (3): 6–15.

Goldstone, Jack A. 2001. "Demography, Environment, and Security." In *Environmental Conflict*, ed. Paul Diehl and Nils Petter Gleditsch, 84–108. Boulder, CO: Westview Press.

Grossmann, Herschel I. 1991. "A General Equilibrium Model of Insurrections." *American Economic Review* 81 (4): 912–21.

Gurr, Ted R. 1970. *Why Men Rebel*. Princeton, NJ: Princeton University Press.

———. 2000. *Peoples Versus States: Minorities at Risk in the New Century*. Washington, DC: United States Institute of Peace Press.

Hauge, Wenche, and Tanja Ellingsen. 1998. "Beyond Environmental Scarcity: Causal Pathways to Conflict." *Journal of Peace Research* 35 (3): 299–317.

Hegre, Håvard, Tanja Ellingsen, Scott Gates, and Nils Petter Gleditsch. 2001. "Toward a Democratic Civil Peace? Democracy, Political Change, and Civil War, 1816–1992." *American Political Science Review* 95 (1): 33–48.

Hegre, Håvard, Ranveig Gissinger, and Nils Petter Gleditsch. 2003. "Globalization and Internal Conflict." In *Globalization and Armed Conflict*, ed. Gerald Schneider, Katherine Barbieri, and Nils Petter Gleditsch, 251–75. Lanham, MD: Rowman and Littlefield.

Hegre, Håvard, and Nicholas Sambanis. 2006. "Sensitivity Analysis of Empirical Results on Civil War Onset." *Journal of Conflict Resolution* 26 (6): 695–715.

Hendrix, Cullen S., and Sarah M. Glaser. 2007. "Trends and Triggers: Climate Change and Civil Conflict in Sub-Saharan Africa." *Political Geography* 26 (6): 695–715.

Homer-Dixon, Thomas F. 1991. "On the Threshold. Environmental Changes as Causes of Acute Conflict." *International Security* 16 (2): 76–116.

———. 1999. *Environment, Scarcity and Violence*. Princeton, NJ: Princeton University Press.

———. 2007. "Terror in the Weather Forecast." *New York Times*, April 24. http://www.nytimes.com/2007/04/24/opinion/24homer-dixon.html?_r=1 &oref=slogin.

IFPRI (International Food Policy Research Institute). 2007. "Annual Report 2006–2007." Washington, DC. http://www.ifpri.org/sites/default/files/publi cations/ar06.pdf.

Kahl, Colin H. 2006. *States, Scarcity, and Civil Strife in the Developing World*. Princeton, NJ: Princeton University Press.

Klare, Michael. 2001. *Resource Wars: The New Landscape of Global Conflict*. New York: Metropolitan Books.

Lacina, Bethany, and Nils Petter Gleditsch. 2005. "Monitoring Trends in Global Combat: A New Dataset of Battle Deaths." *European Journal of Population* 21 (2–3): 145–66.

Levy, Marc A., Catherine Thorkelson, Charles Vörösmarty, Ellen Douglas, and Macartan Humphreys. 2006. "Freshwater Availability Anomalies and Outbreak of Internal War: Results from a Global Spatial Time Series Analysis." Paper prepared for the Human Security and Climate Change Workshop, Holmen, Norway, June 21–23.

Martin, Adrian. 2005. "Environmental Conflict Between Refugee and Host Communities." *Journal of Peace Research* 42 (3): 329–46.

Miguel, Edward, Shanker Satyanath, and Ernest Sergenti. 2004. "Economic Shocks and Civil Conflict: An Instrumental Variables Approach." *Journal of Political Economy* 112 (4): 725–53.

Muller, Edward N., and Erich Weede. 1990. "Cross-National Variation in Political Violence: A Rational Action Approach." *Journal of Conflict Resolution* 34 (4): 624–51.

Murshed, S. Mansoob, and Scott Gates. 2005. "Spatial-Horizontal Inequality and the Maoist Insurgency." *Nepal Review of Development Economics* 9 (1): 121–34.

Nel, Philip, and Marjolein Righarts. 2008. "Natural Disasters and the Risk of Violent Civil Conflict." *International Studies Quarterly* 51 (1): 159–84.

Ohlsson, Leif. 2003. "The Risk of Livelihood Conflicts and the Nature of Policy Measures Required." In *The Future of Peace in the Twenty-First Century*, ed. Nicholas Kittrie, Rodrigo Carazo-Odio, and James Mancham. Washington, DC: Carolina Academic Press.

O'Lear, Shannon, and Paul F. Diehl. 2007. "Not Drawn to Scale: Research on Resource and Environmental Conflict." *Geopolitics* 12 (1): 166–82.

Østby, Gudrun. 2006. "Horizontal Inequalities, Political Environment and Civil Conflict: Evidence from 55 Developing Countries, 1986–2003." Working Paper 28, Centre for Research on Inequality, Human Security and Ethnicity, University of Oxford, UK. http://www.research4development.info/PDF/Outputs/Inequality/ R8230b.pdf.

———. 2008. "Polarization, Horizontal Inequalities and Violent Civil Conflict." *Journal of Peace Research* 45 (2): 143–82.

Østby, Gudrun, Ragnhild Nordås, and Jan Ketil Rød. 2009. "Regional Inequalities and Civil Conflict in Sub-Saharan Africa." *International Studies Quarterly* 53 (2): 301–24.

Parry, Martin, Osvaldo F. Canziani, Jean Palutikof, Paul. van der Linden, and Clair Hanson, eds. 2007. *Climate Change 2007: Impacts, Adaptation and Vulnerability. Contribution of Working Group II to the Fourth Assessment Report of the Intergovernmental Panel on Climate Change.* Cambridge, U.K.: Cambridge University Press.

Peluso, Nancy Lee, and Michael Watts. 2001. "Violent Environments." In *Violent Environments*, ed. Nancy Lee Peluso and Michael Watts, 3–38. Ithaca, NY: Cornell University Press.

Prunier, Gerard. 2007. *Darfur: The Ambiguous Genocide.* Ithaca, NY: Cornell University Press.

Ragin, Charles C. 1987. *The Comparative Method: Moving Beyond Quantitative and Qualitative Strategies.* Los Angeles, CA: University of California Press.

Raleigh, Clionadh, Lisa Jordan, and Idean Salehyan. 2008. "Assessing the Impact of Climate Change on Migration and Conflict." Paper presented at the World Bank workshop on the Social Dimensions of Climate Change, Washington, DC, March 5–6.

Raleigh, Clionadh, and Henrik Urdal. 2007. "Climate Change, Environmental Degradation and Armed Conflict." *Political Geography* 26 (6): 674–94.

Reuveny, Rafael. 2007. "Climate Change-Induced Migration and Conflict." *Political Geography* 26 (6): 656–73.

Reynal-Querol, Marta. 2002. "Ethnicity, Political Systems, and Civil Wars." *Journal of Conflict Resolution* 46 (1): 29–54.

Rosecrance, Richard. 1986. *The Rise of the Trading State: Commerce and Conquest in the Modern World.* New York: Basic Books.

Salehyan, Idean. 2007. "Transnational Rebels: Neighboring States and Sanctuaries for Rebel Groups." *World Politics* 59 (2): 217–42.

———. 2008. "From Climate Change to Conflict? No Consensus Yet." *Journal of Peace Research* 45 (3): 315–26.

Schwartz, Daniel, Tom Deligiannis, and Thomas Homer-Dixon. 2001. "The Environment and Violent Conflict." In *Environmental Conflict,* ed. Paul Diehl and Nils Petter Gleditsch, 273–94. Boulder, CO: Westview Press.

Schwartz, Peter, and Doug Randall. 2003. "An Abrupt Climate Change Scenario and Its Implications for United States National Security: Imagining the Unthinkable." U.S. Department of Defense, Washington, DC.

Small, Christopher, and Robert Nicholls. 2003. "A Global Analysis of Human Settlement in Coastal Zones." *Journal of Coastal Research* 19 (3): 584–99.

Stern, Nicholas. 2006. *The Economics of Climate Change: The Stern Review.* Cambridge, U.K.: Cambridge University Press.

Stewart, Frances. 2000. "Crisis Prevention: Tackling Horizontal Inequalities." *Oxford Development Studies* 28 (3): 245–62.

Stewart, Frances, Graham Brown, and Luca Mancini. 2005. "Why Horizontal Inequalities Matter: Some Implications for Measurement." Working Paper 19, Centre for Research on Inequality, Human Security, and Ethnicity, Oxford, U.K. http://www.crise.ox.ac.uk/pubs.shtml.

Strand, Håvard. 2007. "Reassessing the Democratic Civil Peace." PhD diss., Department of Political Science, University of Oslo, Norway.

Suliman, Mohamed. 1997. "Ethnicity from Perception to Cause of Violent Conflicts: The Case of the Fur and Nuba Conflicts in Western Sudan." Institute for African Alternatives, London. http://www.sudanarchive.net.

————. 1999. "The Rationality and Irrationality of Violence in Sub-Saharan Africa." In *Ecology, Politics and Violent Conflict*, ed. Mohamed Suliman, 27–43. London: Zed.

Theisen, Ole Magnus. 2008. "Blood and Soil? Resource Scarcity and Internal Armed Conflict Revisited." *Journal of Peace Research* 45 (6): 801–18.

UNEP (United Nations Environment Programme). 2007. "Environmental Degradation Triggering Tensions and Conflict in Sudan." UNEP, Nairobi, Kenya. http://www.unep.org/Documents.Multilingual/Default.asp?DocumentID=512 &ArticleID=5621&l=en.

UCDP/PRIO (Uppsala Conflict Data Program/International Peace Research Institute of Oslo). 2008. Armed Conflict Dataset. http://www.prio.no/cscw/datasets.

Urdal, Henrik. 2005. "People vs. Malthus: Population Pressure, Environmental Degradation, and Armed Conflict Revisited." *Journal of Peace Research* 42 (4): 417–34.

Vreeland, James Raymond. 2008. "The Effect of Political Regime on Civil War: Unpacking Anocracy." *Journal of Conflict Resolution* 52 (3): 401–25.

Watts, Jonathan. 2007. "Riots and Hunger Feared as Demand for Grain Sends Food Costs Soaring." *The Guardian*, December 4. http://www.guardian.co.uk/world/2007/dec/04/china.business.

CHAPTER 4

Climate Change and Migration: Emerging Patterns in the Developing World

Clionadh Raleigh and Lisa Jordan

Climate change is expected to bring about significant changes in migration patterns throughout the developing world. This chapter addresses the "environmental refugees" concept. We argue that this term conflates the idea of disaster victim with that of refugee, and reduces the complexity of real situations. The environmental security literature often presents climate change as an external push factor to which migration is the mechanical response; but speculation about the social consequences of climate change has relied on worst-case scenarios and broad generalizations concerning the links between physical processes and social consequences (see Myers 1993, 2002 and Döös 1997).

Here we review the relevant literature on the environmental drivers of migration. To address directly how the environment influences migration, we review, discuss, and observe how previous environmental disasters influenced mobility. We focus on chronic disasters, sudden-onset disasters, and climate extremes because, as climate researchers have emphasized, an increase in the frequency and severity of such events is the most likely short-term to medium-term effect of climate changes:[1] "Recently, it has become more evident that climate change will not express itself primarily through slow shifts in average temperature over a long period [...] there is mounting evidence that it is extreme events, such as droughts, floods and heat waves that we must prepare for" (Helmer and Hilhorst 2006, p. 1; van Aalst 2006; IPCC 2001). Chronic disasters include droughts, degradation, and desertification; sudden-onset disasters are floods, tsunamis, hurricanes, and other swift weather events. Climatic extremes are defined as those bringing about more permanent changes, such as sea level rise and increasing temperatures.

We roundly conclude that large-scale community relocation resulting from either chronic or sudden-onset environmental disasters related to climate change is unlikely to be a common response over the next 20 years. Where we do find such large-scale relocations, it is likely to reflect policy and response failures of public institutions more than the specific nature of the hazard. This conclusion is built on the four major findings from the environmental migration literature:

1. Disasters vary considerably in their potential to instigate migration.
2. Individuals and communities in the developing world incorporate environmental risk into their livelihoods, contingent on their available assets. Adaptations to increased environmental change and economic assets, social position, political relationships, and government policies shape variation.
3. During periods of chronic environmental degradation, the most common response by individuals and communities is to intensify labor migration patterns.
4. With the onset of a sudden disaster or the continued presence of a chronic disaster (that is, drought or famine), communities engage in distress migration patterns characterized by short-term relocations to nearby areas.

Nevertheless, the concept of the environmental refugee remains popular, regardless of its deviation from both the practices of those most affected by climate changes and variations and its dismissal of the well-researched multicausal nature of migration patterns within the developing and developed worlds. In summary, the relationship between environmental changes and migration is complicated, but the primacy or independent weight ascribed to the environment is largely unsubstantiated. Certainly, the patterns of migration predicted by proponents of the environmental refugee framework are largely at odds with the actual migratory processes that multiple case studies and subnational research have recorded and repeatedly confirm. Indeed, the main issue with the environmental refugee concept is its underestimation of the historical patterns, categories, causes, and processes of adaptation and migration within high-risk environments.

We conclude that the available evidence does support a cautious approach to the association between climate change and migration in the short term. By extending the time frame into future generations, issues surrounding climatic thresholds, coping strategies, and cumulative disasters become critical factors not yet fully considered in migration literature. Perhaps of most significance, conclusions from previous cases verify that uneven development and governance policies have exacerbated the depth of disaster impacts and shaped the choices available to those populations affected by environmental disasters. We expect that these factors will continue to exert the strongest influence on those people who are adversely affected by climate changes.

Climate Change Risk, Vulnerability, and Adaptation

The Intergovernmental Panel on Climate Change initially gave the following warning: "The gravest effects of climate change may be those on human migration as millions are displaced by shoreline erosion, coastal flooding and severe drought" (IPCC 1990, p. 20). Since 1990, there have been significant changes in the panel's position as it has recognized that a variety of complex interactions mediate migratory decision making. Subsequent reports adopted more nuanced depictions of migration, primarily by redirecting the focus in terms of "human vulnerability" (IPCC 2001). In fact, reference to *human* migration as a *consequence* of climate change was eliminated from the 2001 "Summary for Policy Makers" (IPCC 2001). This major shift relates to how climate change risk has been reconceptualized. Specifically, the current framework for examining the social consequences of climate change recognizes that physical vulnerability constitutes only one factor in a person's overall vulnerability to environmental hazards. Vulnerability is a concept used to describe the relative risk of adverse changes in the environment as it is experienced by individuals, households, and communities. It is a construction based on the ability to anticipate, cope with, resist, and recover from a disaster (Adger 2000). The economic, political, and social vulnerabilities of the individual, the community, and the nation make up the overall risk involved in climate-related changes; and they complicate the concept of vulnerability considerably. (For discussions of vulnerability, see Smit and Wandel 2006, Blaikie et al. 2004, Adger 2000, and Smit et al. 2000.)

Multiple models of vulnerability have been advanced recently in disaster literature, but vulnerability assessments are not associated with widely accepted indicators or methods of measurement (McLeman and Smit 2006; Ringius et al. 1997). However, a measure of vulnerability *must* incorporate the sensitivity of the system, the nature of exposure, and the capacity of the people exposed to prepare and cope with risks (see Füssel 2007 and Smit and Wandel 2006). We contend that the vulnerability of an individual or community within the developing world in sustaining a livelihood in the face of climate change and variation can be understood best through a scalar approach. It is built on "everyday issues," such as livelihoods and marginal social status, which may contribute to poor land management practices, resource pressures, and increasing reliance on degraded resources. These are compounded by "episodic issues," such as flooding or droughts (Bryant and Bailey 2003). The distribution of costs involved in everyday and episodic changes is not random. The poor and otherwise marginalized members of society are disproportionately affected by all disasters.[2] The main point of this approach is to emphasize that we live in a "politicized environment" where the costs and benefits associated with environmental change are distributed unequally among actors (Bryant and Bailey 2003).

To be vulnerable to climate changes, however, does not make someone a potential "climate migrant." Ecological hazards occur with sufficient frequency to influence how people incorporate such risks into their livelihoods (McLeman and Smit 2006). People in marginal regions have developed a great variety of mechanisms to strengthen their ability to cope with both slow climatic changes and extreme climatic events (Meze-Hausken 2000; Mula 1999; Maxwell 1996; Findley 1994). Discussions of climate change coping mechanisms typically are located at the household level, and a number of broad conclusions are evident from case study literature (Henry 2006; McLeman and Smit 2006). How a household reacts to environmental hazards depends on the severity of the change, the household's particular vulnerabilities, and the assets and strategies available to it (Meze-Hausken 2000; Mortimore 1989).

Multiple factors unrelated to environmental change influence adaptation most directly. The availability of markets, access to infrastructure, and the promise and delivery of aid influence the ability of families to prepare for and withstand environmental hazards and changes (Eriksen, Brown, and Kelly 2005). Although factors such as war, government controls on movement, and employment opportunities are often beyond the control of families and communities, they strongly shape actions and movements in response to calamities.

Communities experiencing chronic environmental hazards generally mitigate risk through livelihood diversification. Rural livelihoods typically are composed of a combination of three strategies: agropastoral activities, livelihood diversification, and migration (de Haan, Brock, and Coulibaly 2002). Typical labor migration is a critical component of rural livelihoods because migrant wages provide investment capital for rural commodity production and the experience of migration is a conduit for the flow of new ideas and social practices into rural areas (Baker and Aida 1995). A severe stress situation, such as drought, brings into stark focus the ways in which diverse income sources and dynamic coping strategies form the basis of rural livelihoods (Eriksen, Brown, and Kelly 2005). During such times, coping strategies tend to become more specialized and directed toward surviving droughts and insulating families against "distress migration" (Eriksen, Brown, and Kelly 2005; McGregor 1994). To a great extent, vulnerability depends on the ability of individuals to specialize successfully. Although coping strategies tend to contract during nondrought periods, the maintenance of indigenous coping institutions is found to be crucial for continued existence in marginal lands (Eriksen, Brown, and Kelly 2005; McCabe 1990).

People in areas prone to sudden-onset disasters have a range of coping strategies that are largely based on their available assets and social networks. In wealthier states, insurance against destruction caused by disasters is common for households in flood plains, in fire-prone areas, and on fault lines. In developing states, coping mechanisms and social networks are tied closely, indicating that losses resulting from disaster will be shared among those in a community or group. International migration is an important household strategy for risk reduction because it has been shown that remittances greatly reduce vulnerability in recovering from disasters (Suleri and Savage 2007; Young 2006). With regard to climatic extremes, many small island societies have proved to be resilient in the face of past social and environmental upheaval (Bayliss-Smith et al. 1988). Resilience is based on traditional knowledge, institutions and technologies, opportunities for migration and remittances, land tenure regimes, the subsistence economy, and the links between state and customary decision making (Barnett 2001; Barnett and Adger 2003).

To summarize, ecological calamities have occurred with sufficient frequency to influence how people incorporate such risks into their livelihoods (McLeman and Smit 2006). The three most critical strategies when living on degraded land or in uncertain ecological climates are diversification of livelihood, consolidation of savings into incontestable forms, and social investment (that is, migration). Short-term labor migration is an important household strategy for risk reduction because remittances greatly reduce people's vulnerability when recovering from disasters (Suleri and Savage 2007; Young 2006). Initial assets and networks underscore coping strategies. Migration is only one of a variety of survival strategies pursued by families, either simultaneously or consecutively with other coping strategies (McLeman and Smit 2006; de Bruijn and van Dijk 2003; Reardon 1997; McGregor 1994; Painter, Price, and Sumberg 1994; and Cleveland 1991). When hazards and climate changes become so severe and common that they destroy the abilities of households and communities to mediate their situations and risks, distress migration or massive livelihood changes are posited to occur.

Links between Migration and Climate Change

Early work on environmental migration has also presumed that physical vulnerability to a disaster occupied a primary role in affecting mobility patterns. Proponents of the concept of the environmental refugee—such as El-Hinnawi (1985), Hugo (1996), and Myers (1993, 2002)—accept certain premises: that international destinations have increasing significance (with the concession that most migration is within countries); that migration, as both a cause and a consequence of environmental change, occurs in poor countries; and that the scale and pace of environmental change have accelerated, so environmentally induced migration also is increasing. Such premises allow for "predictive estimates," which suggest that numbers may range from 10 million (Jacobson 1988, p. 6) to 150 million (Myers 1993, p. 200), to the most recent total of 1 billion (Christian Aid 2007, p. 1). The validity of these estimates can be questioned because they are based mainly on conjecture and worst-case scenarios.

A brief analysis of the numbers of people affected by various disasters counters perceptions of high disaster-migration rates. The following tables and discussion rely on our summaries of data compiled by the Centre for Research on the Epidemiology of Disasters (CRED 2008). Table 4.1 displays results of descriptive statistics that are consistent with several assertions advanced within disaster literature.[3] Specifically, chronic environmental hazards (such as drought) are not the most common, but they do affect the most people (at an average of 10 percent of a country's population). In low-income states, the effect is heightened to 13 percent of a country's population. The range of drought-affected populations is significant, with multiple cases (8) reaching an affected rate above 90 percent.

Table 4.1. Consequences of Environmental Hazards across Countries

Disaster	Number of Reported Events[b]	All Countries			Low-Income Countries[a]		
		Population Affected[c]	Population Killed[d]	Population Homeless[e]	Population Affected	Population Killed	Population Homeless
Drought	Overall: 332 Low-income: 261	10% SD 21.000 (0–100%)	<1% SD 0.700 (0–1%)	<1% SD 0.020 (0–0.4%)	13% SD 22.00 (0–100%)	<1% SD 0.080 (0–1%)	<1% SD 0.020 (0–0.42%)
Extreme temperature	Overall: 324 Low-income: 148	<1% SD 2.500 (0–40%)	<1% SD 0.600 (0–0.05%)	<1% SD 0.002 (0–0.05%)	<1% SD 3.39 (0–40%)	<1% SD 0.010 (0–0.01%)	<1% SD 0.002 (0–0.05%)
Flood	Overall: 2,839 Low-income: 1801	<1% SD 3.000 (0–48%)	<1% SD 0.020 (0–0.12%)	<1% SD 0.500 (0–27%)	<1% SD 3.64 (0–48%)	<1% SD 0.010 (0–0.038%)	<1% SD 0.700 (0–27%)
Landslides	Overall: 451 Low-income: 311	<1% SD .014 (0–2.5%)	<1% SD 0.004 (0–0.1%)	<1% SD 0.800 (0–2%)	<1% SD 0.17 (0–2.5%)	<1% SD 0.005 (0–0.09%)	<1% SD 0.800 (0–2%)
Wave/surge	Overall: 34 Low-income: 25	<1% SD 1.200 (0–6%)	<1% SD 0.030 (0–1%)	<1% SD 0.700 (0–4%)	<1% SD 0.50 (0–3%)	<1% SD 0.030 (0–0.18%)	<1% SD 0.500 (0–2.5%)
Wind storms	Overall: 2,311 Low-income: 519	1.1% SD 7.500 (0–100%)	<1% SD 0.010 (0–0.42%)	<1% SD 4.480 (0–100%)	2% SD 10.00 (0–100%)	<1% SD 0.020 (0–0.42%)	<1% SD 6.310 (0–100%)

Source: Authors' compilation, using CRED (2008).

Note: SD = standard deviation.

a. Low-income countries are those with an annual GDP per capita of less than $3,000.

b. Number of reported events is the number of Emergency Events Database entries for countries from 1970 to 2007.

c. Affected people are those requiring immediate assistance during a period of emergency—that is, requiring basic survival needs such as food, water, shelter, sanitation, and immediate medical assistance (includes the appearance of a significant number of cases of an infectious disease introduced in a region or a population that is usually free from that disease).

d. The population killed includes persons confirmed dead and those presumed dead.

e. The homeless population includes people needing immediate assistance with shelter.

This is most clear in East and West Africa, where affected populations reach 14 percent and 22 percent, respectively (table 4.2). Southern Asia has many more people affected (more than 32 million), at a mean rate of 11 percent of the population. Only in the case of drought is a significant proportion of a state affected; and in those cases, there is little more than the presumption of migration. As discussed below, the actual patterns of migration in both chronically degraded areas and those affected by swift disasters are, for the most part, internal, temporary, and short term.

Still other national studies attempt to discern the "weight" that can be ascribed to the environment when determining migration rates and patterns. Barrios, Bertinelli, and Strobl (2006) have found that, although rainfall variation is correlated with increased rates of urbanization on the sub-Saharan continent, there is no evidence that this applies to the rest of the developing world. They add that climatic change in itself is seldom the direct root of migration (except in extreme cases of flood and drought). In response, other studies emphasized the role of nonenvironmental drivers and variation, such as social policies, within the developing world (Duranton 2008).

In short, there is no established methodology for determining patterns of migration resulting from environmental changes (see Kniveton et al. 2008). However limited, the evidence suggests that, in certain circumstances, environmental hazards do alter the migration patterns typically observed in developing countries. The literature on sudden, natural disaster–induced migration does not support the notion that massive and ceaseless migration flows will follow disasters. National-level studies allude to these relationships but, because of severe data limitations on migration, these studies are unable to provide a compelling narrative about mobility as a result of environmental change. Instead, within the disaster literature there is a clear distinction made between where and what is affected, the coping mechanisms of those who stay in a disaster area, the migration patterns of those induced to flee, and the return process of forced migrants. Overall, wholesale community relocation in reaction to natural disasters is a relatively rare occurrence (Hunter 2005).

Climate Impacts and Migration Decisions

Much of the practical information about environmental migration can be gleaned from dozens of case studies of both chronic and swift-onset disasters. The findings across most studies are remarkably consistent.

Table 4.2. Population Affected by Select Disasters across Global Subregions

Region	Subregion[a]	Droughts[b]	Extreme Temperatures	Floods	Slides	Wave/Surges	Wind Storms
Americas	Caribbean (283)	268,636 (12%)	n.a.	42,304 (1%)	512 (1%)	n.a.	104,241 (5%)
	North (612)	30,000 (<1%)	200 (0%)	200,035 (<1%)	1,531 (<1%)	n.a.	5,000,047 (2%)
	Central (356)	58,933 (2%)	1052 (<1%)	26,198 (<1%)	708 (<1%)	1,720 (<1%)	103,808 (2%)
	South (599)	1,905,980 (7%)	131,927 (<1%)	136,544 (<1%)	7,425 (<1%)	931 (<1%)	15,545 (<1%)
Africa	East (401)	1,765,088 (14%)	n.a.	108,167 (2%)	562 (<1%)	27,556 (2%)	118,167 (3%)
	Middle (88)	374,726 (9%)	n.a.	25,990 (<1%)	73 (<1%)	n.a.	9,645 (<1%)
	North (145)	1,700,243 (7%)	40 (<1%)	98,628 (<1%)	3323 (<1%)	12 (<1%)	24,402 (<1%)
	South (94)	295,531 (15%)	21 (<1%)	24,111 (1%)	34 (<1%)	n.a.	48,314 (4%)
	West (200)	967,841 (22%)	333,359 (13%)	52,944 (<1%)	519 (<1%)	n.a.	4,822 (<1%)
Asia	Central (65)	n.a.	200,008 (1.5%)	33,735 (<1%)	3,502 (<1%)	n.a.	2,505 (<1%)
	Eastern (856)	9,934,389 (1%)	3,132 (<1%)	6,413,745 (1%)	1,580 (<1%)	9,693 (<1%)	999,417 (<1%)
	South East (864)	974,805 (7%)	n.a.	258,548 (1%)	10,490 (<1%)	64,640 (<1%)	369,193 (<1%)
	Southern (1,051)	32,600,000 (11%)	5,248 (<1%)	2,461,976 (1%)	58,129 (<1%)	294,222 (2%)	423,754 (<1%)
	Western (209)	302,900 (6%)	652 (<1%)	57,770 (1%)	240 (<1%)	n.a.	4,293 (<1%)
Europe	East (288)	0 (0%)	14,508 (<1%)	49,474 (<1%)	281 (<1%)	n.a.	48,356 (1%)
	North (103)	n.a.	37 (<1%)	n.a.	38 (<1%)	n.a.	n.a.
	Russian Federation (46)	n.a.	n.a.	6,084 (<1%)	1,411 (<1%)	n.a.	2,610 (<1%)
	South (270)	1,023,333 (13%)	1,417 (<1%)	26,601 (<1%)	1,262 (<1%)	n.a.	12,904 (<1%)
	West (221)	n.a.	1,406 (<1%)	5,646 (<1%)	715 (<1%)	13 (<1%)	39,868 (<1%)
Oceania	Australia/New Zealand (197)	1,011,429 (6%)	920,161 (5%)	1,556 (<1%)	243 (<1%)	n.a.	40,662 (<1%)
	Melanesia (126)	139,149 (8%)	n.a.	25,830 (2%)	2,029 (<1%)	6096 (<1%)	18,336 (4%)
	Micronesia (18)	56,400 (5%)	n.a.	n.a.	n.a.	n.a.	1,334 (1%)
	Polynesia (46)	n.a.	n.a.	4 (<1%)	178 (<1%)	n.a.	11,213 (15%)

Source: Authors' compilation, using CRED (2008).

Note: n.a. = not applicable. In parentheses are the percents by region and disaster, as a proportion of the national population during the year of the disaster.
a. These are Emergency Events Database (EM-DAT)–designated subregions, with the total number of EM-DAT disaster entries in parentheses.
b. Each disaster total is the total number of affected people (including those killed and homeless), by subregion.

Temporary, circular, and internal migration is an important aspect of spatial mobility in drought-affected areas; and is often a component of household survival strategies for coping with drought and high levels of production uncertainty (Henry, Schoumaker, and Beauchemin 2004; Roncoli and Ingram 2001; Hampshire and Randall 1999; Guilmoto 1998; Hill 1990; Reardon, Matlon, and Delgado 1988). Much of the migration that occurs in areas of drought or increasing degradation is believed to be economically motivated. The destinations of these migrants are often proximate urban areas or, increasingly, other rural areas within a state.

However, sudden-onset disasters and prolonged chronic hazards lead to "distress migration." These migrations are composed largely of impoverished people seeking aid until they may be able to return to their homes or communities (if that is possible). Two characteristics of distress-induced migration deserve emphasis: The first characteristic is forced migration as a result of ecological disaster, which mainly results in internal rather than international displacement (see de Haan 2002, Findley 1994, and McGregor 1994). The second characteristic is the temporary rather than permanent nature of the displacement caused by such stressors. Living mainly in poor countries, victims have little mobility and few options (Lonergan 1998); and the majority of displaced people return as soon as possible to reconstruct their homes in the disaster zone (Naik, Stigter, and Laczko 2007; Kliot 2004). Results confirm this with remarkable regularity (Piguet 2008). If permanent migration is the result of a disaster, it is seen as a reflection of the state's deficient response rather than of the natural hazard's impact (Oliver-Smith 2004; Castles 2002; Black 2001; Wood 2001).

Chronic Disaster Migration. Drought caused by physical and climate changes is a significant cause of livelihood insecurity. Declines in the ability of households to be self-sustaining are related to climatic vagaries, long-term declines in production (that is, degradation), increasing population growth, and land shortages. However, the exposure and risk of households and communities differ significantly as a function of marginalization, land tenure arrangements, coping strategies, opportunities and market infrastructure, and the availability of government assistance. A link between environmental conditions and migration has been measured in several settings in rural Africa.[4] Migration patterns resulting from chronic drought conditions initially follow preestablished labor migration patterns and may not differ in intensity from areas with established high rates of temporary, circular migration (Henry, Boyle, and Lambin 2003;

Findley 1994). Migrations within rural areas are still the dominant type of internal migration flow, and they involve both short-distance and long-distance moves. The former disproportionately involve women migrating for family reasons, and the latter for economic reasons (Henry, Schoumaker, and Beauchemin 2004).

Migration from and to rural areas intensifies following a major drought or a poor harvest as a way to minimize risk (Ezra 2001; Pederson 1995; Findley 1994). These migratory patterns are generally circular. They are especially popular among poor (but not necessarily the poorest) people who may not have resources to instigate migration. Henry, Schoumaker, and Beauchemin (2004) note that people from drier regions are more likely than those from wetter areas to engage in both temporary and permanent migrations to other rural areas. Rainfall deficits tend to increase the risk of long-term migration to rural areas and decrease the risk of short-term moves to distant locations. Because the ability to migrate is contingent on socioeconomic situations, migrant numbers may decline in times of decreasing rainfall.[5] Contrary to expectations, international long-term migration occurs following high-yield seasons because families and communities can invest in such movements (Deshingkar 2006; Henry, Schoumaker, and Beauchemin 2004; and Kuhn 2000).

The literature points to differences in migrant composition flows and destinations over time and across countries. Massey, Axinn, and Ghimire (2007) find that, in Nepal, long-distance moves are predicated on perceived declines in land productivity, but the effect is weaker than for short-distance mobility. This effect, however, is confined only to lower and non-Hindu castes. No environmental characteristics appear to affect the odds of making a distant move, thus casting doubt on the utility of the concept of the environmental refugee in explaining interregional or international migration. Migration in response to drought was found in only 2 percent of households in areas of India and Bangladesh during 1983 and 1994/95 (Paul 1995, table 3).[6] In these instances, increased migration was not a response to drought conditions, partially because substantial labor migration had taken place previously (Paul 1995). Most people depend on remittances from such labor migrants or family networks to continue living in drought-affected areas (Henry, Schoumaker, and Beauchemin 2004; Caldwell, Reddy, and Caldwell 1986). Limited assets and government policies were the determining factors (Findley 1994). However, increases in rural-rural migration as a way to diversify income sources may increase during periods of drought (Henry, Schoumaker, and Beauchemin 2004;

de Haan 1999, 2000, 2002). In comparison to other disasters where few victims consider permanently changing location, however, the percentage of people considering migration was highest in drought-affected areas (ranging from 10 percent to 31 percent) (Perch-Nielsen 2004, p. 81; Burton, Kates, and White 1993).[7]

Fast-Onset Migration. The characteristics of distress migration differ within and across countries, depending on the severity of a crisis, the ability of a household to respond, the geography of the crisis, evacuation opportunities, existing and perpetuating vulnerabilities, available relief, and intervening government policies. Case studies confirm that household and community responses to disasters are shaped primarily by compensation opportunities, income restoration possibilities, and community support over relocation and resettlement possibilities (Colson 2003; Turton 2003). After a brief period, displacees and forced migrants return to their home areas at a remarkably high rate (exceeding 90 percent) (Perch-Nielsen, Bättig, and Imboden 2008, p. 381; Suhrke 1994; Berry and Downing 1993; Belcher and Bates 1983).

Across studies of fast-onset disaster responses, there is a general consensus that those who are displaced locally rely heavily on social capital, community networks, and economic resources to structure decisions. Distance to possible hosting areas is a crucial factor for distress migrants because people often make proximate moves (Paul 2005; Perch-Nielsen 2004). The number of people seeking relief aid varies, depending on the geography of relief, infrastructure, instability, predisaster assets, and past experiences with aid distribution (Ezra 2001; Ezra and Kiros 2001; McGregor 1994). The geography of relief is a critical factor: In very severe emergencies, urban areas can be popular destinations for forced migrants. Research on Kenyan and Somali reactions to climate hazards and drought conditions notes a swelling of population around market towns as a result of a growing dependence on both aid and markets for a sustainable lifestyle (Little et al. 2001). Regional urban pushes have been found in similar contexts: "Migrants who can do so commonly head for towns and cities and famines swell peri-urban shantytowns with new arrivals. Those who are far from towns and lead agrarian lifestyles may suffer more from famines" (Shipton 1990, p. 353).

In areas of frequent and compounded natural disasters, short migrations are found to be common and a component of coping within an ecologically marginal area. In studies of riverbank erosion in Bangladesh, Haque and Zaman (1989) find that in one of the worst-affected subdistricts, 60 percent of residents had been displaced at least once in their lifetimes, and that 98 percent had moved fewer than 5 miles. Zaman and Wiest (1991) found that people moved an average of 2 miles from their previous residences because there was a persistent belief that it was critical to stay close to family and that land would be reclaimed (also see Hutton and Haque 2004). Multiple displacements are common characteristics of Bangladeshi charland settlements.[8] Although most people return to reestablish their livelihoods when new land subsequently reemerges, a considerable proportion of displacees (10–25 percent) move to urban centers and become permanent squatter-settlers. These urban migrants cited economic factors, including landlessness, poverty, unemployment, and natural hazards as the major causes of the rural push (Islam 1996).

These patterns are also evident among distress migrants from developed states. In their analyses of migration patterns following hurricanes in the United States in 1996 and 2004, Smith and McCarty (1996, 2006) find that 12–30 percent of people initially displaced left their homes permanently. Of those migrants, most stayed in the same county or state (61–73 percent stayed within the county, 9–23 percent within the state, and 15–18 percent of those who permanently resettled left the state) (Smith and McCarty 2006, p. 6).[9] Hurricane Katrina is believed to have created more permanent migrants than did previous disasters in the area because it was far more destructive, led to massive infrastructure damage (especially among people who were uninsured), and devastated the local economy—leading to a significant rise in unemployment (Perch-Nielsen, Bättig, and Imboden 2008; Smith and McCarty 2006).

Thus, a summary of results on the migration choices of displaced victims of natural disasters confirms, with rare exceptions, a strong propensity to stay in local areas or to return (Piguet 2008; Burton, Kates, and White 1993).

Climatic Extremes. A special case is the effect of sea-level rise and erosion on migration potential. Indeed, previous evidence of riverbank erosion in Bangladesh did lead to some sizable migrations (Mahmood 1995; Zaman 1989). Future projections of sea-level rise call for a consideration of resettlement as an adaptive strategy to climate change, particularly in very high-risk countries, such as the Pacific Islands and low-lying atolls (Barnett 2001). The limited research that has been undertaken supports opposite claims regarding the ability of sea-level rise to displace coastal populations.

For example, through an historical study of an island in the Chesapeake Bay, Arenstam Gibbons and Nicholls (2006) find high levels of migration took place before the area became physically uninhabitable, largely because of a decrease in community and social services in the area. The resettlement decision was predicated chiefly on nonenvironmental issues. The authors conclude that this historic example largely conforms to Barnett and Adger's (2003) concept of "socioecological thresholds," at which social issues surrounding extremes like sea-level rise may shape responses more than direct physical impacts shape them.

However, considerable resilience to short-term hazards has been documented in the Pacific Islands (Campbell 1990; Marshall 1979; Lessa 1964; Rappaport 1963; Firth 1959). "Sufficient evidence exists to show that people have maintained habitation of the Pacific Islands during periods of substantial exogenous and human induced environmental changes, although adaptation was at times traumatic" (Barnett 2001, p. 986). This may be the result of cross-island community efforts in times of need, such as after a cyclone, when communities would assist each other through the redistribution of food or would allow for the dispersal of people to other islands. More recently, smaller-scale migrations within home islands were observed in Samoa and Tokelau during Cyclone Ofa (Campbell 1998; Hooper 1990). This requires good social relationships with "neighbors" and increased cooperation at the regional level (Nicholls and Mimura 1998; Torry 1979). There is some concern that those island links that did exist have been weakened and replaced by connections with more distant countries because remittances now constitute a large proportion of postdisaster assistance (Campbell 1998). As populations on each island decrease because of labor out-migration, the predisaster resilience of those people who remain also is strengthened by remittances. As with other disasters, people have integrated a variety of different adaptations into their livelihoods (Perch-Nielsen 2004; Black 2001).

"Managed retreat" or the progressive abandonment of land and structures in highly vulnerable areas and resettlement of inhabitants is mentioned frequently in reference to erosion and sea-level rise. To date, however, no such movements have taken place. Retreat may be an option for sparsely inhabited coasts, but is unlikely in urban areas (Perch-Nielsen 2004; Leatherman 2001). Permanent resettlement of high-risk populations in disaster zones is now considered a possible strategy to address environmental problems, particularly sea-level rise in low-lying or overpopulated island-states. For the most part, government-induced resettlement has a very poor reputation as a response to development, conflict, or environmental problems. This reputation is mainly the result of inadequate planning and facilities, the politicized nature of the resettlement process, and the general inability of governments to address postresettlement issues. In cases of both voluntary and involuntary movements, governments face the same three issues: people are hesitant to move, there are considerable settlement and development issues in new locations, and people often attempt to return to their home areas (see Cernea 1997; Chan 1995; and Smith 2004).

Although resettlement may be successful in reducing people's physical vulnerability to disaster risk, it is often coupled with a decrease in development and living standards, thereby possibly increasing the economic and social vulnerability of resettled populations. This results mainly from issues surrounding employment, land acquisition, water resources, migrants' unequal access to resources and opportunities, and a decrease in social networks and capital (Badri et al. 2006).

Summary of the Relationship between Natural Disasters and Migration. This overview of climate change–related migration has focused on chronic and sudden-onset disaster areas. Case studies confirm that a short-term to medium-term increase and intensification of typical labor migration should be expected out of degraded and drought/famine areas, whereas initial local displacement will characterize movements from sudden-onset disaster areas. A small share of migrants may choose to relocate permanently. (Case studies cited above have noted a range of 0–30 percent.) No mass migrations should or are expected to occur (Hunter 2005). These findings deviate substantially from more egregious estimates from the environmental refugee literature, mainly because such discussions fail to take into account human reaction and adaptation to change (Black 2001).

The available literature and this chapter have stressed the role of predisaster coping and resilience strategies designed to address household and community vulnerability to, and risk of, environmental hazards. Distressed migrants experience subnational socioeconomic impoverishment and marginalization as a consequence of involuntary migration. This is partly a socially constructed process, reflecting inequitable access to land and other resources (Hutton and Haque 2004; Blaikie et al. 2004). The majority of urban displacees endure cumulative and increasing impoverishment, and has limited opportunities to relieve debt and attain savings that might ease the hardships associated with displacement (Haque 1997; Greenberg and Schneider 1996). In extremely severe cases, large-scale distress migration can be accompanied by "abject misery, large-scale beggary and greatly increased mortality" (Adhana 1991, p. 187).

Effects of Government Policies on Environmental Migrants

Political, economic, social, environmental, and household factors affect vulnerability and adaptation to environmental hazards. Therefore, it follows that policies that influence vulnerabilities will affect the production of migrants as a result of climate change (Hunter 2005). Previous and current attempts by governments to address environmental vulnerabilities bear out three main conclusions:

1. Multiple efforts, including microcredit lending for sustainable (environmentally conscious) development and improvement of livelihoods, food security in poorer countries with semiarid climates, improved planning of coastal communities, and fair trade programs are not "climate change programs" in themselves; but they do act to reduce the negative impacts of migrations that are directly or indirectly the result of climate change. Furthermore, government restrictions on internal labor migration limit its use as "insurance" against depleted rural livelihoods (see UNFPA 2007, Deshingkar 2006, de Bruijn and van Dijk 2003, and de Haan 2002).

2. The ways in which governments respond to disasters are predicated largely on the kind of political relationships that existed between sectors before the crisis (Pelling and Dill 2006). Multiple case studies support this assertion, specifically in reference to famines, disaster relief, and postdisaster assistance (Paul 2005; de Bruijn and van Dijk 2003; de Waal 1997; McGregor 1994; Bern et al. 1993; Berry and Downing 1993; Cutler 1993).

3. Political instability, economic crises, and uneven development can lower the threshold for coping with environmental disaster. These issues increase mortality and morbidity, and generally lead to longer and more sustained political and environmental crises. This issue of "compounded disasters" is especially relevant because both political and environmental crises are more common within the developing world (Raleigh and Urdal 2007; Lischer 2005; Johnson 2003; de Waal 1997; Suhrke 1993; Zolberg, Suhrke, and Aguayo 1989).

In conclusion, although it is difficult to gauge properly the impact of government policies on environmentally induced migration, it seems clear that development policies—independent of climate change policy—strongly shape the risks of communities in disaster-prone regions. Governments can bolster a community's immunity to disaster by encouraging local and urban development, thereby lessening its social and economic vulnerability to hazards. A range of other polices designed to reduce physical risks and increase adaptation are not widespread.

Future Research in Thresholds and Compounded Disasters

We conclude our review of the environmental migration literature with a discussion of future challenges. As the pace of climate change and variability increases, individuals and communities may face lower thresholds for environmental disasters, cumulative disasters, and increased competition for resource access. We contend that future research should focus on these facets for environmental security, and discuss each in turn.

Thresholds

Coping mechanisms are an ever-important component of sustainable rural livelihoods in the developing world. Overall, people with more survival strategies can resist migration longer. However, coping mechanisms are exhausted quickly during periods of subsequent or multiple disasters. Although distress migration patterns are curtailed by migrant remittances (allowing people to survive long-term food shortages), no definitive link between the level of vulnerability and the time of relocation has been detected during especially severe cases. Tipping points of social, institutional, and environmental vulnerabilities may lessen a community's ability to be resilient (Smit and Wandel 2006). This might signify a type of threshold or limit within the society with respect to climate, making all people in a region similarly negatively affected by drought, independent of their initial entitlements and household situations. Such a threshold is exhibited in "chronically vulnerable areas" where coping mechanisms are stretched (Meze-Hausken 2000; Findley 1994). More research should be undertaken in areas where the margin for disaster is exceedingly narrow.

Presently, there is little information about (1) varying thresholds, (2) the short- and long-term actions of people from a region past its threshold, and (3) the socioeconomic conditions that influence thresholds. The conclusions of Little et al. (2001) note the presence of distress migrants in peri-urban settlements in consistently degraded areas of the Horn of Africa. Hence, a rapid increase in the rate of urbanization is possible, although without the present cyclical and circular patterns. In those situations within the developing world, competition for employment and resources can become violent or, at the very least, politicized along ethnic lines (Raleigh and Hegre 2009; Raleigh and Urdal 2007; Urdal 2005; Goldstone 2002). However, there is contrary evidence that finds that new migrants across borders or in refugee camps tend to "employ a myriad of strategies which include the redefinition of kinship and social obligations" (Giuffrida 2005, p. 538; Harrell-Bond, Voutira, and Leopold 1992). Attempts to bridge ethnic gaps clearly are a priority to migrants, who most likely will not engage in conflict far from a solid support base. No evidence exists connecting labor migrants or disaster refugees to increased conflict.

Cumulative Disasters and Complex Emergencies

Disaster situations within already unstable areas can lead to complex emergencies. These emergencies are characterized by the breakdown or failure of state structures, intercommunal violence, disputed legitimacy of authority, the potential for assistance to be misused, abuse of human rights, and the deliberate targeting of civilian populations by violent forces. Research into the dynamics of complex emergencies is woefully underdeveloped. A way forward may be to observe the propensity for violence and rebellion in chronically vulnerable areas, while isolating underlying causes and trigger mechanisms (Homer-Dixon 1991, 1994).

People affected by adverse climatic changes are challenged by sociopolitical instability in multiple ways: (1) economically marginalized and politically excluded people may reside in areas less developed than the rest of the state; (2) disasters often hit politically peripheral regions hardest, catalyzing regional political tensions; and (3) existing inequalities can be exacerbated by postdisaster governmental manipulation (Raleigh 2009; Suhrke 1994; Zolberg, Suhrke, and Aguayo 1989).

The onset of an Ethiopian food crisis in 2001 demonstrates the danger of cumulative vulnerability to environmental hazards (Hammond and Maxwell 2002). The combined effects of agricultural shocks in 1997, advanced droughts in the southern border regions, conflict in Eritrea, and the migration of several thousand pastoralists into Ethiopia from southwestern Somalia and northern Kenya created an untenable situation. Eventual delivery of relief to the most affected regions helped minimize distress migration. The onset of cumulative disasters is difficult to assess because of the specifics and contexts present in individual cases. In general, contingent factors (such as drought) are monitored; but critical underlying factors (such as household assets and variable destitution levels) are not integrated into disaster prediction models.

Although the potential for civil war resulting from environmental variation is limited (see Raleigh and Urdal 2007), a rise in structural violence may occur as disaster victims continue to exist in marginal rural and urban areas. Conflict dynamics within the developing world are least likely to involve the most destitute people; instead, regional and ethnic elites are most likely to engage over state and environmental resources (see Peluso and Watts 2001). Those groups appear to be the least likely to be proportionally affected by environmental disasters. In short, we contend that environmental migration should be considered a development concern rather than a future security issue. Indeed, in the places most likely to be adversely affected by climate change and variability, development is aligned closely with both environment and security matters. However, the development approach considers how people are active agents in incorporating environmental change into their livelihoods; a security or environmental refugee approach is concerned primarily with the external effects of people affected by environmental change. Such a perspective is not based on the coping and migration processes and patterns evident within the developing world.

Conclusion

As we have demonstrated, the adaptation of people adversely affected by climate changes will be based on predisaster characteristics, economic opportunities, and political stability. The effect of internal and regional development in shaping the choices available to migrants cannot be overstated. Furthermore, the politicization of development and relief is a critical component in understanding future challenges in the environmental security nexus.

Notes

1. In the past decade, weather-related natural hazards have been the cause of 90 percent of natural disasters and 60 percent of related deaths (IFRC 2005).

The effects are especially dire in developing countries, where environmental hazard victims represented 98 percent of all disaster-affected populations (IFRC 2005).

2. The Maasai people of Kenya present an appropriate example of the interaction between physical and social vulnerabilities. They are considered marginalized because their access to social services, infrastructure, and political representation routinely is well below national averages in remote and low-population-density pastoral areas (Coast 2002). If drought should affect Maasai and non-Maasai territory, Maasai people would be most vulnerable to severe and crippling economic effects because their margin for disaster is constructed so narrowly by forces partially beyond their control.

3. CRED is responsible for compiling the Emergency Events Database (EM-DAT; http://www.emdat.be/). EM-DAT does not record the number of migrating victims as a result of disasters, but does provide the number of people affected, killed, or made homeless as a result. Affected people are those who require immediate assistance during a period of emergency—that is, those with such basic survival needs as food, water, shelter, sanitation, and immediate medical care (including the appearance of a significant number of cases of an infectious disease introduced in a region or a population that usually is free from that disease). For more information, see the "Explanatory Notes" tab at the Web site.

4. As examples, see Henry, Schoumaker, and Beauchemin (2004) for Burkino Faso; see Ezra (2001) for Ethiopia; and see Findley (1994) for Mali.

5. For Burkina Faso, see Henry, Schoumaker, and Beauchemin (2004); for Ethiopia, see Meze-Hausken (2004); for Mexico, see Black et al. (2008) and Kniveton et al. (2008); for Mali, see Findley (1994); and for Sudan, see Haug (2002).

6. Estimates on out-migration during the latter drought differ. Paul (1995; cited also in Smith 2004, p. 244) finds that during the 1994/95 drought in Bangladesh, only 1 of the 265 households surveyed (or 0 percent), had to resort to migration.

7. Additional case studies on drought include McLeman (2006); Henry, Boyle, and Lambin (2003); Mahran (1995); Paul (1995); Autier et al. (1989); Corbett (1988); Caldwell, Reddy, and Caldwell (1986); Warrick (1980); and Prothero (1968).

8. In 1995, the Flood Plan Coordination Organization (1995) estimated that 728,000 people were displaced between 1981 and 1993. More than 40 percent of the displaced squatters had been uprooted three or four times, and 36 percent had been displaced 5–10 times. Another 14 percent had been displaced more than 10 times (Hutton and Haque 2004, p. 46).

9. Eighteen percent of people leaving the state following the 2004 hurricanes amounted to 35,000 people. Smith and McCarty (2006, p. 10) and Perch-Nielsen, Bättig, and Imboden (2008) note that this number should be considered in tandem with the 300,000–400,000 people leaving Florida annually.

References

Adger, W. Neil. 2000. "Institutional Adaptation to Environmental Risk under the Transition in Vietnam." *Annals of the Association of American Geographers* 90 (4): 738–58.

Adhana, Adhana Haile. 1991. "Peasant Response to Famine in Ethiopia, 1975–1985." *Ambio* 20 (5): 186–88.

Arenstam Gibbons, Sheila J., and Robert J. Nicholls. 2006. "Island Abandonment and Sea-Level Rise: An Historical Analog from the Chesapeake Bay, USA." *Global Environmental Change* 16 (1): 40–47.

Autier, Philippe, Jean-Pierre d'Altilia, Bart Callewaert, Baalti Tamboura, Jean-Pierre Delamalle, and Vincent Vercruysse. 1989. "Migrations and Nutritional Status in the Sahel." *Disasters* 13 (3): 247–54.

Badri, Seyed Ali, Ali Asgary, A. R. Eftekhari, and Jason Levy. 2006. "Post-Disaster Resettlement, Development and Change: A Case Study of the 1990 Manjil Earthquake in Iran." *Disasters* 30 (4): 451–68.

Baker, Jonathan, and Tade Akin Aida, eds. 1995. *The Migration Experience in Africa.* Uppsala, Sweden: Nordiska Afrikainstitutet.

Barnett, Jon. 2001. "Adapting to Climate Change in Pacific Island Countries: The Problem of Uncertainty." *World Development* 29 (6): 977–93.

Barnett, Jon, and W. Neil Adger. 2003. "Climate Dangers and Atoll Countries." *Climatic Change* 61: 321–37.

Barrios, Salvador, Luisito Bertinelli, and Eric Strobl. 2006. "Climatic Change and Rural-Urban Migration: The Case of Sub-Saharan Africa." *Journal of Urban Economics* 60 (3): 357–71.

Bayliss-Smith, Tim P., Richard Bedford, Harold Brookfield, and Marc Latham. 1988. *Islands, Islanders and the World: The Colonial and Post-Colonial Experience of Eastern Fiji.* Cambridge, U.K.: Cambridge University Press.

Belcher, John C., and Frederick L. Bates. 1983. "Aftermath of Natural Disasters: Coping through Residential Mobility." *Disasters* 7 (2): 118–28.

Bern, C., J. Sniezek, G. M. Mathbor, M. S. Siddiqi, C. Ronsmans, A. M. Chowdhury, A. E. Choudhury, K. Islam, M. Bennish, E. Noji, and R. I. Glass. 1993. "Risk Factors for Mortality in the Bangladesh Cyclone of 1991." *Bulletin of the World Health Organization* 71 (1): 73–78. http://whqlibdoc.who.int/bulletin/1993/Vol71-No1/bulletin_1993_71(1)_73-78.pdf.

Berry, Leonard, and Thomas E. Downing. 1993. "Drought and Famine in Africa, 1981–86: A Comparison of Impacts and Responses in Six Countries." In *The Challenge of Famine: Recent Experience, Lessons Learned,* ed. John Osgood Field, 35–58. West Hartford, CT: Kumarian Press.

Black, Richard. 2001. "Environmental Refugees: Myth or Reality?" New Issues in Refugee Research Working Paper 34, United Nations High Commissioner for Refugees, Geneva, Switzerland.

Black, Richard, Dominic Kniveton, Ronald Skeldon, Daniel Coppard, Akira Murata, and Kerstin Schmidt-Verkerk. 2008. "Demographics and Climate Change: Future Trends and Their Policy Implications for Migration." Working Paper T-27, Development Research Centre on Migration, Globalisation and Poverty, University of Sussex, Brighton, U.K.

Blaikie, Piers, Terry Cannon, Ian Davis, and Ben Wisner. 2004. *At Risk: Natural Hazards, People's Vulnerability and Disasters.* 2nd ed. London: Routledge.

Bryant, Raymond L., and Sinéad Bailey. 2003. *Third World Political Ecology.* 2nd ed. London: Routledge.

Burton, Ian, Robert W. Kates, and Gilbert F. White. 1993. *The Environment as Hazard.* 2nd ed. New York: Guilford Press.

Caldwell, John C., P. H. Reddy, and Pat Caldwell. 1986. "Periodic High Risk as a Cause of Fertility Decline in a Changing Rural Environment: Survival Strategies in the 1980–1983 South Indian Drought." *Economic Development and Cultural Change* 34 (4): 677–701.

Campbell, John. 1990. "Disaster and Development in Historical Context: Tropical Cyclone Response in the Banks Islands, Northern Vanuatu." *International Journal of Mass Emergencies and Disasters* 8 (3): 401–24

———. 1998. "Consolidating Mutual Assistance in Disaster Management within the Pacific: Principles and Application." Presentation to the Seventh South Pacific Regional IDNDR Disaster Management Meeting, Nadi, Fiji, September 23–25.

Castles, Steven. 2002. "Environmental Change and Forced Migration: Making Sense of the Debate." Working Paper 70, United Nations High Commissioner for Refugees, Geneva, Switzerland.

Cernea, Michael M. 1997. "The Risk and Reconstruction Model for Resettling Displaced Populations." *World Development* 25 (10): 1569–88.

Chan, Ngai Weng. 1995. "Flood Disaster Management in Malaysia: An Evaluation of the Effectiveness of Government Resettlement Schemes." *Disaster Prevention and Management* 4 (4): 22–29.

Christian Aid. 2007. *Human Tide: The Real Migration Crisis. A Christian Aid Report.* London: Christian Aid.

Cleveland, David. 1991. "Migration in West Africa: A Savannah Village Perspective." *Africa* 61 (2): 222–46.

Coast, Ernestina. 2002. "Maasai Socio-Economic Conditions: Cross Border Comparison." *Human Ecology* 30 (1): 79–105.

Colson, Elizabeth. 2003. "Forced Migration and the Anthropological Response." *Journal of Refugee Studies* 16 (1): 1–18.

Corbett, Jane. 1988. "Famine and Household Strategies." *World Development* 16 (9): 1099–112.

CRED (Centre for Research on the Epidemiology of Disasters). 2008. "EM-DAT: Emergency Events Database." Université Catholique de Louvain, Belgium. http://www.emdat.be/.

Cutler, Peter. 1993. "Responses to Famine: Why They Are Allowed to Happen." In *The Challenge of Famine: Recent Experience, Lessons Learned*, ed. John Osgood Field, 72–87. West Hartford, CT: Kumarian Press.

de Bruijn, Mirjam, and Han van Dijk. 2003. "Changing Population Mobility in West Africa: Fulbe Pastoralists in Central and South Mali." *African Affairs* 102: 285–307.

de Haan, Arjan. 1999. "Livelihoods and Poverty: The Role of Migration. A Critical Review of the Migration Literature." *Journal of Development Studies* 36 (2): 1–47.

———. 2000. "Migrants, Livelihoods and Rights: The Relevance of Migration in Development Policies." Social Development Working Paper 4, U.K. Department for International Development, London.

———. 2002. "Migration and Livelihoods in Historical Perspective: A Case Study of Bihar, India." *Journal of Development Studies* 38 (5): 115–42.

de Haan, Arjan, Karen Brock, and Ngolo Coulibaly. 2002. "Migration, Livelihoods and Institutions: Contrasting Patterns of Migration in Mali." *Journal of Development Studies* 38 (5): 37–58.

Deshingkar, Priya. 2006. "Internal Migration, Poverty and Development in Asia." Briefing Paper 11, Overseas Development Institute, London.

de Waal, Alex. 1997. *Famine Crimes: Politics and the Disaster Relief Industry in Africa*. African Issues Series. Oxford, U.K.: James Currey.

Döös, Bo R. 1997. "Can Large-Scale Environmental Migrations Be Predicted?" *Global Environmental Change* 7 (1): 41–61.

Duranton, Giles. 2008. "Viewpoint: From Cities to Productivity and Growth in Developing Countries." *Canadian Journal of Economics* 41 (3): 689–736.

El-Hinnawi, Essam. 1985. *Environmental Refugees*. New York: United Nations Development Programme.

Eriksen, Siri, Katrina Brown, and P. Mitch Kelly. 2005. "The Dynamics of Vulnerability: Locating Coping Strategies in Kenya and Tanzania." *Geographical Journal* 171 (4): 287–305.

Ezra, Markos. 2001. "Demographic Responses to Environmental Stress in the Drought- and Famine-Prone Areas of Northern Ethiopia." *International Journal of Population Geography* 7: 259–79.

Ezra, Markos, and Gebre-Egziabher Kiros. 2001. "Rural Out-Migration in the Drought-Prone Areas of Ethiopia: A Multilevel Analysis." *International Migration Review* 35 (3): 749–71.

Findley, Sally. 1994. "Does Drought Increase Migration? A Study of Migration from Rural Mali during the 1983–1985 Droughts." *International Migration Review* 28 (3): 539–53.

Firth, Raymond. 1959. *Social Change in Tikopia*. New York: Macmillan.

Flood Plan Coordination Organization. 1995. "Final Report on the Flood Action Plan." Government of Bangladesh, Dhaka.

Füssel, Hans-Martin. 2007. "Vulnerability: A Generally Applicable Conceptual Framework for Climate Change Research." *Global Environment Change* 17: 155–67.

Giuffrida, Alessandra. 2005. "Clerics, Rebels and Refugees: Mobility Strategies and Networks among the Kel Antessar." *Journal of North African Studies* 10 (3–4): 529–43.

Goldstone, Jack A. 2002. "Population and Security: How Demographic Change Can Lead to Violent Conflict." *Journal of International Affairs* 56 (1): 3–21.

Greenberg, Michael, and Dona Schneider. 1996. *Environmentally Devastated Neighbourhoods: Perceptions, Policies, and Realties.* New Brunswick, NJ: Rutgers University Press.

Guilmoto, Christopher Z. 1998. "Institutions and Migrations: Short-Term Versus Long-Term Moves in Rural West Africa." *Population Studies* 52 (1): 85–103.

Hammond, Laura, and Daniel Maxwell. 2002. "The Ethiopian Crisis of 1999–2000: Lessons Learned, Questions Unanswered." *Disasters* 26 (3): 262–79.

Hampshire, Kate, and Sara Randall. 1999. "Seasonal Labor Migration Strategies in the Sahel: Coping with Poverty or Optimising Security." *International Journal of Population Geography* 5 (5): 367–85.

Haque, C. Emdad. 1997. *Hazards in a Fickle Environment: Bangladesh.* Dordrecht, Netherlands: Kluwer Academic.

Haque, C. Emdad, and Muhammad Q. Zaman. 1989. "Coping with Riverbank Erosion Hazard and Displacement in Bangladesh: Survival Strategies and Adjustments." *Disasters* 13 (4): 300–14.

Harrell-Bond, Barbara, Eftihia Voutira, and Mark Leopold. 1992. "Counting the Refugees: Gifts, Givers, Patrons and Clients." *Journal of Refugee Studies* 5 (3–4): 205–25.

Haug, Ruth. 2002. "Forced Migration, Processes of Return and Livelihood Construction among Pastoralists in Northern Sudan." *Disasters* 26 (1): 70–84.

Helmer, Madeleen, and Dorothea Hilhorst. 2006. "Natural Disasters and Climate Change." *Disasters* 30 (1): 1–4.

Henry, Sabine. 2006. "Some Questions on the Migration-Environment Relationship." Panel presentation to the Population-Environment Research Network Cyberseminar on Rural Household Micro-Demographics, Livelihoods, and the Environment, April.

Henry, Sabine, Paul Boyle, and Eric F. Lambin. 2003. "Modelling Inter-Provincial Migration in Burkina Faso, West Africa: The Role of Socio-Demographic and Environmental Factors." *Applied Geography* 23 (2–3): 115–36.

Henry, Sabine, Bruno Schoumaker, and Cris Beauchemin. 2004. "The Impact of Rainfall on the First Out-Migration: A Multi-Level Event-History Analysis in Burkino Faso." *Population and Environment* 25 (5): 423–60.

Hill, Allen. 1990. "Demographic Responses to Food Shortages in the Sahel." *Population and Development Review* 15 (Suppl): 168–92.

Homer-Dixon, Thomas F. 1991. "On the Threshold: Environmental Changes as Causes of Acute Conflict." *International Security* 16 (2): 76–116.

———. 1994. "Environmental Scarcities and Violent Conflict: Evidence from Cases." *International Security* 19 (1): 5–40.

Hooper, Andrew. 1990. "Tokelau." In *Climatic Change: Impacts on New Zealand*, ed. Ministry for the Environment, 210–14. Wellington, NZ: Ministry for the Environment.

Hugo, Graham. 1996. "Environmental Concerns and International Migration." *International Migration Review* 30 (1): 105–31.

Hunter, Lori M. 2005. "Migration and Environmental Hazards." *Population and Environment* 26 (4): 273–302.

Hutton, David, and C. Emdad Haque. 2004. "Human Vulnerability, Dislocation and Resettlement: Adaptation Processes of River-Bank Erosion-Induced Displacees in Bangladesh." *Disasters* 28 (1): 41–62.

IFRC (International Federation of Red Cross and Red Crescent Societies). 2005. *World Disasters Report 2005: Focus on Information in Disasters*. Geneva, Switzerland: IFRC.

IPCC (Intergovernmental Panel on Climate Change). 1990. "Policymakers' Summary of the Potential Impacts of Climate Change." Report from Working Group II to the Intergovernmental Panel on Climate Change, Australia.

———. 2001. *Climate Change 2001: Synthesis Report*. Cambridge, U.K.: Cambridge University Press.

Islam, Nazrul. 1996. "Sustainability Issues in Urban Housing in a Low-Income Country: Bangladesh." *Habitat International* 20 (3): 377–88.

Jacobson, Jodi I. 1988. "Environmental Refugees: A Yardstick of Habitability." Worldwatch Paper 86, Worldwatch Institute, Washington, DC.

Johnson, Douglas Hamilton. 2003. *The Root Causes of Sudan's Civil Wars*. African Issues Series. Oxford, U.K.: James Currey.

Kliot, Nurit. 2004. "Environmentally Induced Population Movements: Their Complex Sources and Consequences—A Critical Review." In *Environmental Change and Its Implications for Population Migration*, ed. Jon D. Unruh, Maarten S. Krol, and Nurit Kliot. Dordrecht, Netherlands: Kluwer Academic.

Kniveton, Dominic, Kerstin Schmidt-Verkerk, Christopher Smith, and Richard Black. 2008. *Climate Change and Migration: Improving Methodologies to Estimate Flows*. Report 33. Geneva, Switzerland: International Organization for Migration.

Kuhn, Randall. 2000. "The Logic of Letting Go: Family and Individual Migration from Bangladesh." Paper presented at BRAC, Dhaka, Bangladesh.

Leatherman, Stephen P. 2001. "Social and Economic Costs of Sea Level Rise." In *Sea Level Rise: History and Consequences*, ed. Bruce C. Douglas, Michael S. Kearney, and Stephen P. Leatherman, 181–223. San Diego, CA: Academic Press.

Lessa, William. 1964. "The Social Effects of Typhoon Ophelia on Ulithi." *Micronesia* 1: 1–47.

Lischer, Sarah Kenyon. 2005. *Dangerous Sanctuaries: Refugee Camps, Civil War, and the Dilemmas of Humanitarian Aid*. Ithaca, NY: Cornell University Press.

Little, Peter D., Kevin Smith, Barbara A. Cellarius, D. Layne Coppock, and Christopher Barrett. 2001. "Avoiding Disaster: Diversification and Risk Management among East African Herders." *Development and Change* 32 (3): 401–33.

Lonergan, Stephen. 1998. "The Role of Environmental Degradation in Population Displacement." In *Global Environmental Change and Human Security*. Report 4, 5–15. Victoria, BC: University of Victoria.

Mahmood, Raisul. 1995. "Emigration Dynamics in Bangladesh." *International Migration* 33 (3-4): 699–728.

Mahran, Hatim. 1995. "The Displaced, Food Production and Food Aid." In *War and Drought in Sudan: Essays on Population Displacement*, ed. Eltigani El Tigani, 63–74. Gainesville, FL: University Press of Florida.

Marshall, Mac. 1979. "Natural and Unnatural Disaster in the Mortlock Islands of Micronesia." *Human Organization* 38 (3): 265–72.

Massey, Douglas S., William Axinn, and Dirgha Ghimire. 2007. "Environmental Change and Out-Migration: Evidence from Nepal." Research Report 07-615, Population Studies Center, University of Michigan, Ann Arbor.

Maxwell, Simon. 1996. "Food Security: A Post-Modern Perspective." *Food Policy* 21 (2): 155–70.

McCabe, J. Terrence. 1990. "Success and Failure: The Breakdown of Traditional Drought Coping Institutions among the Pastoral Turkana of Kenya." *Journal of Asian and African Studies* 25 (3-4): 146–60.

McGregor, JoAnn. 1994. "Climate Change and Involuntary Migration: Implications for Food Security." *Food Policy* 19 (2): 120–32.

McLeman, Robert. 2006. "Migration Out of 1930s Rural Eastern Oklahoma: Insights for Climate Change Research." *Great Plains Quarterly* 26 (1): 27–40.

McLeman, Robert, and Barry Smit. 2006. "Migration as an Adaptation to Climate Change." *Climatic Change* 76 (1–2): 31–53.

Meze-Hausken, Elisabeth. 2000. "Migration Caused by Climate Change: How Vulnerable Are People in Dryland Areas?" *Mitigation and Adaptation Strategies for Global Change* 5 (4): 379–406.

———. 2004. "Contrasting Climate Variability and Meteorological Drought with Perceived Drought and Climate Change in Northern Ethiopia." *Climate Research* 27 (1): 19–31.

Mortimore, Michael. 1989. *Adapting to Drought: Farmers, Famines, and Desertification in West Africa*. Cambridge, U.K.: Cambridge University Press.

Mula, Rosana. 1999. "Coping with Mother Nature: Household Livelihood Security and Coping Strategies in a Situation of a Continuing Disaster in Tarlac, Philippines." PhD diss., Landbouwuniversiteit, Wageningen, Netherlands.

Myers, Norman. 1993. *Ultimate Security: The Environmental Basis of Political Stability.* New York: Norton.

———. 2002. "Environmental Refugees: A Growing Phenomenon of the 21st Century." *Philosophical Transactions of the Royal Society: Biological Sciences* 357: 609–13.

Naik, Asmita, Elca Stigter, and Frank Lazcko. 2007. *Migration, Development and Natural Disasters: Insights from the Indian Ocean Tsunami.* Migration Research Series. International Organization for Migration, Geneva, Switzerland.

Nicholls, Robert J., and Nobuo Mimura. 1998. "Regional Issues Raised by Sea-Level Rise and Their Policy Implications." *Climate Research* 11: 5–18.

Oliver-Smith, Anthony. 2004. "Theorizing Vulnerability in a Globalized World, A Political Ecological Perspective." In *Mapping Vulnerability: Disasters, Development and People,* ed. Greg Bankoff, Georg Frerks, and Dorothea Hilhorst, 10–24. London: Earthscan.

Painter, Thomas, Thomas Price, and James Sumberg. 1994. "Your *Terroir* and My 'Action Space': Implications of Differentiation, Mobility, and Diversification for the *Approche Terroir* in Sahelian West Africa." *Africa* 64 (4): 447–64.

Paul, Bimal Kanti. 1995. "Farmers and Public Responses to the 1994–95 Drought in Bangladesh: A Case Study." Quick Response Report 76, Natural Hazards Research and Applications Information Center, Boulder, CO. http://www .colorado.edu/hazards/research/qr/qr76.html.

——— 2005. "Evidence against Disaster-Induced Migration: The 2004 Tornado in North-Central Bangladesh." *Disasters* 29 (4): 370–85.

Pedersen, Jon. 1995. "Drought, Migration and Population Growth in the Sahel— The Case of the Malian Gourma, 1900–1991." *Population Studies* 49 (1): 111–26.

Pelling, Mark, and Kathleen Dill. 2006. "'Natural' Disasters as Catalysts of Political Action." Briefing Paper 06/01, Chatham House, Royal Institute of International Affairs, London.

Peluso, Nancy Lee, and Michael Watts, eds. 2001. *Violent Environments.* Ithaca, NY: Cornell University Press.

Perch-Nielsen, Sabine L. 2004. "Understanding the Effect of Climate Change on Human Migration: The Contribution of Mathematical and Conceptual Models." Master's thesis, Department of Environmental Studies, Swiss Federal Institute of Technology, Zurich, Switzerland.

Perch-Nielsen, Sabine L., Michele B. Bättig, and Dieter Imboden. 2008. "Exploring the Link between Climate Change and Migration." *Climatic Change* 91 (3–4): 375–93.

Piguet, Etienne. 2008. "Climate Change and Forced Migration." New Issues in Refugee Research Working Paper 153, United Nations High Commissioner for Refugees, Geneva, Switzerland.

Prothero, R Mansell. 1968. "Migration in Tropical Africa." In *The Population of Tropical Africa*, ed. John Charles Caldwell and Chukula Okonjo, 250–60. London: Longman.

Raleigh, Clionadh. 2009. "Patterns of Conflict across Ethnic Communities in Central Africa." Unpublished manuscript, Trinity College Dublin, Ireland.

Raleigh, Clionadh, and Håvard Hegre. 2009. "Population Size, Concentration, and Civil War: A Geographically Disaggregated Analysis." *Political Geography* 28 (3): 224–38.

Raleigh, Clionadh, and Henrik Urdal. 2007. "Climate Change, Environmental Degradation and Armed Conflict." *Political Geography* 26 (6): 674–94.

Rappaport, Roy. 1963. "Aspects of Man's Influence upon Island Ecosystems: Alteration and Control." In *Man's Place in the Island Ecosystem*, ed. F. Fosberg, 155–70. Honolulu, HI: Bishop Muesum Press.

Reardon, Thomas. 1997. "Using Evidence of Household Income Diversification to Inform Study of the Rural Nonfarm Labour Market in Africa." *World Development* 25 (5): 735–47.

Reardon, Thomas, Peter Matlon, and Christopher Delgado. 1988. "Coping with Household Level Food Security in Drought Affected Areas of Burkino Faso." *World Development* 16 (9): 1065–74.

Ringius, Lasse, Tom Downing, Mike Hulme, and Dominic Waughray. 1997. "Adapting to Climate Change in Africa: Prospects and Guidelines." *Mitigation and Adaptation Strategies for Global Change* 2 (1): 19–44.

Roncoli, Carla, and Keith Ingram. 2001. "The Costs and Risks of Coping with Drought: Livelihood Impacts and Farmers' Responses in Burkino Faso." *Climate Research* 19: 119–32.

Shipton, Parker. 1990. "African Famines and Food Security." *Annual Review of Anthropology* 13: 353–94.

Smit, Barry, Ian Burton, Richard J. T. Klein, and Johanna Wandel. 2000. "An Anatomy of Adaptation to Climate Change and Variability." *Climatic Change* 45 (1): 223–51.

Smit, Barry, and Johanna Wandel. 2006. "Adaptation, Adaptive Capacity and Vulnerability." *Global Environmental Change* 16: 282–92.

Smith, Kevin. 2004. *Environmental Hazards: Assessing Risk and Reducing Disasters*. London: Routledge, 4th edition.

Smith, Stanley K., and Christopher McCarty. 1996. "Demographic Effects of Natural Disasters: A Case Study of Hurricane Andrew." *Demography* 33 (2): 265–75.

———. 2006. "Florida's 2004 Hurricane Season: Demographic Response and Recovery." Paper presented at the Annual Meeting of the Southern Demographic Association, Durham, NC, November 3.

Suhrke, Astri. 1993. "Pressure Points: Environmental Degradation, Migration and Conflict." Paper prepared for the workshop on Environmental Change, Population Displacement, and Acute Conflict, Ottawa, Canada, June 1994.

———. 1994. "Environmental Degradation and Population Flows." *Journal of International Affairs* 47: 473–96.

Suleri, Abid Qaiyum, and Kevin Savage. 2007. "Remittances in Crises: A Case Study from Pakistan." Humanitarian Policy Group Background Paper, Overseas Development Institute, London.

Torry, William I. 1979. "Anthropological Studies in Hazardous Environments: Past Trends and New Horizons." *Current Anthropology* 20 (3): 517–40.

Turton, David. 2003. "Refugees and 'Other Forced Migrants': Towards a Unitary Study of Forced Migration." Paper prepared for the Workshop on Settlement and Resettlement in Ethiopia, Addis Ababa, January 28–30.

UNFPA (United Nations Population Fund). 2007. *State of World Population 2007: Unleashing the Potential of Urban Growth.* New York: UNFPA.

Urdal, Henrik. 2005. "Defusing the Population Bomb: Is Security a Rationale for Reducing Global Population Growth?" *Environmental Change and Security Program Report* 11: 5–11.

van Aalst, Maarten K. 2006. "The Impacts of Climate Change on the Risk of Natural Disasters." *Disasters* 30 (1): 5–18.

Warrick, Richard. 1980. "Drought in the Great Plains: A Case Study of Research on Climate and Society in the USA." In *Climatic Constraints and Human Activities*, ed. Jesse H. Ausubel and Asti K. Biswas, 93–123. Oxford, U.K.: Pergamon Press.

Wood, William B. 2001. "Ecomigration: Linkages between Environmental Change and Migration." In *Global Migrants, Global Refugees: Problems and Solutions*, ed. Aristide R. Zolberg and Peter M. Benda, 42–61. Oxford, U.K.: Berghahn Books.

Young, Helen. 2006. "Livelihoods, Migration and Remittance Flows in Times of Crisis and Conflict: Case Studies for Darfur, Sudan." Humanitarian Working Group Background Paper, Overseas Development Institute, London.

Zaman, Muhammad Q. 1989. "The Social and Political Context of Adjustment to Riverbank Erosion Hazard and Population Resettlement in Bangladesh." *Human Organization* 48 (3): 196–205.

Zaman, Muhammad Q., and Raymond E. Wiest. 1991. "Riverbank Erosion and Population Resettlement in Bangladesh." *Practicing Anthropology* 13 (3): 29–33.

Zolberg, Aristide R., Astri Suhrke, and Sergio Aguayo. 1989. *Escape from Violence: Conflict and the Refugee Crisis in the Developing World.* Oxford, U.K.: Oxford University Press.

CHAPTER 5

The Gender Dimensions of Poverty and Climate Change Adaptation

Justina Demetriades and Emily Esplen

It is generally recognized that people who are already poor and marginalized experience the impacts of climate change most acutely (see Tanner and Mitchell 2008; and GTZ, cited in Lambrou and Piana 2006), and are in the greatest need of adaptation strategies in the face of shifts in weather patterns and resulting environmental phenomena. At the same time, those who are poor and marginalized have the least capacity or opportunity to prepare for the impacts of a changing climate, or to participate in national and international negotiations on tackling climate change.

Also, it is generally recognized that pervasive gender inequalities in societies throughout the world give rise to higher rates of poverty among women, relative to men, and to a more severe experience of poverty by women than by men. This is true for female household heads and among women and girls living within male-headed households considered to be nonpoor as a result of unequal intrahousehold distribution of power and resources, such as food and property (Kabeer 2008; Chant 2007). Where women and girls have less access to and control over resources (material, financial, and human), and have fewer capabilities than men, these impediments undermine their capacity to adapt to existing and predicted impacts of climate change, and to contribute important knowledge and insights to adaptation and mitigation decision-making processes. The situation is exacerbated by the fact that climate change also reinforces *existing* gender inequalities in the key dimensions that are most crucial for coping with climate-related

---

This chapter draws heavily on Brody, Demetriades, and Esplen (2008).

change, including inequalities in access to wealth, new technologies, education, information, and other resources such as land.

Drawing on the limited existing body of literature on gender and climate change, this chapter presents some of the most commonly articulated links between gender inequality, poverty and other forms of exclusion, and vulnerability/resilience to environmental stress and shocks. It focuses in particular on two areas that feature especially prominently in the literature on gender and climate change: (1) agricultural production and livelihoods and (2) climate change-related disasters and conflict. The chapter discusses gendered impacts and highlights innovative efforts to produce climate change responses. In the light of entrenched gender inequalities in decision making at all levels, it goes on to discuss the importance of promoting women's and girls' meaningful participation in decision making on climate change, and provides examples of organizations that are putting initiatives in place to achieve such participation. Finally, it reflects critically on the limitations of existing analyses and approaches to gender and climate change—limitations that we believe must be addressed urgently if we are to develop appropriate, holistic climate change responses grounded in, and relevant to, women's and men's lived realities.

Gendered Impacts: Climate Change, Agricultural Production, and Livelihoods

For households dependent on agriculture, land is the most important productive asset (World Bank 2007). But statutory and/or customary laws restrict women's land rights in many parts of the world, which in turn can make it difficult for women farmers to access credit. Research suggests that women receive less than 10 percent of the credit granted to small farmers in Africa (Nair, Kirbat, and Sexton 2004). Without credit, they cannot buy the crucial inputs needed to adapt to environmental stress—new varieties of plant types and animal breeds intended for higher drought or heat tolerance, and new agricultural technologies (APF/NEPAD 2007). These obstacles are exacerbated further by gender biases in institutions that often reproduce assumptions that men are the farmers (Gurung et al. 2006). As a result, agricultural extension services and technologies rarely are available to women farmers (Lambrou and Piana 2006). Without access to land, credit, and agricultural technologies, women farmers face major constraints in their capacity to diversify into alternative livelihoods.

These obstacles can be particularly problematic for households headed by women who cannot rely on male household members to purchase crucial inputs. Moreover, the number of female-headed households often increases when livelihoods are in jeopardy and men outmigrate for work. Women become de facto heads of households and take on men's farming roles in addition to existing agricultural and domestic responsibilities (Laudazi 2003). It may be difficult for a household that is treated as female headed in a husband's absence to retain control over land and other productive assets because of restrictions on women's property and land rights—heightening women's vulnerability at exactly the point at which their responsibilities increase.

Gendered Impacts: Climate Change-Related Disasters and Conflict

Gender inequality is also a major factor contributing to vulnerability in disaster situations, such as Hurricanes Mitch and Katrina or flooding in South and East Asia, that increasingly are being linked to climate change. Women and girls may be particularly vulnerable because of differences in socialization by which girls are not equipped with the same skills as their brothers (skills such as swimming), or because of restrictions on female mobility. For example, it has been well documented that women in Bangladesh did not leave their houses during floods because of cultural constraints on female mobility, and that those who did were unable to swim (Röhr 2006). During Hurricane Mitch, by contrast, more men than women died. It has been suggested that this result arose from existing gender norms in which ideas about masculinity encouraged risky, "heroic" action in a disaster (Röhr 2006).

Gender inequalities also may be exacerbated in the aftermath of disasters. For example, there is some evidence to suggest that women and girls are more likely to become victims of domestic and sexual violence after a disaster, particularly when families have been displaced and are living in overcrowded emergency or transitional housing where they lack privacy. Adolescent girls report especially high levels of sexual harassment and abuse in the aftermath of disasters, and they complain of the lack of privacy they encounter in emergency shelters (Bartlett 2008). The increase in violence often is attributed partly to stress caused by decreased economic opportunities in the period following a disaster, compounded by longer-term unemployment or threatened livelihoods.

It also is well recognized that climate change will—and already is—resulting in a growing scarcity of natural resources, such as water and arable land, in some parts of the world. With heightened competition over diminishing and unequally distributed resources, conflict over resources is set to increase (Röhr 2008; Reuveny 2007; Hemmati 2005). Although currently there is little research explicitly linking climate change with both conflict and gender, there is a considerable body of work on gender and conflict that points to the different roles women and men play in conflicts, and to the differential impacts of conflict on men and women. In Darfur, sexual violence against women and girls (and less documented against boys and men)—occurring in villages, in and around refugee and internally displaced persons camps, and outside camps at times when scarce fuel and water are being collected—provides a stark example of one of the likely gendered effects of increased climate change–related conflicts.

Gendering Climate Change Responses

The examples described above offer only a brief and partial insight into the ways in which gender inequalities (especially when exacerbated by poverty and other forms of inequality) can leave women and girls at a particular disadvantage in coping with the adverse impacts of climate change. Therefore, integration of a gender-sensitive perspective in all climate change responses is essential if policies and programs are to be genuinely responsive to the particular needs and priorities of all stakeholders—women as well as men.

In La Masica, Honduras, for example, there were no reported fatalities after Hurricane Mitch—in part because a disaster agency having provided gender-sensitive training and having involved women and men equally in hazard management activities, led to a quick evacuation when the hurricane struck (Aguilar n.d.). In the aftermath of disasters, helpful responses may involve working with women and girls in emergency shelters and camps to protect them from heightened levels of violence. Such efforts could involve lighting the way to the toilets or finding people who are willing to monitor the route or accompany children, adolescent girls, and women. It also might involve finding ways to ensure their privacy while they are bathing or dressing (Bartlett 2008).

Another innovative area of existing work relates to the engendering of conflict early-warning systems. The United Nations Development Fund for Women has created a set of gender-sensitive early-warning indicators that include increased gender-based violence, increased unemployment among male youths, reduced trust between ethnic groups, and a reduction in women's involvement over land disputes (Moser 2007). Many of these indicators reflect the projected effects of climate change on communities—particularly where resources are being depleted.

Innovative responses also are in place at the community level, spearheaded by women and men who already have a great deal of knowledge of and experience in coping with the impacts of climate change in their own specific contexts. Such responses emerged clearly in a participatory research project conducted by ActionAid International and the Institute of Development Studies with women in rural communities in the Ganga River Basin in Bangladesh, India, and Nepal. The women who took part in the research described various adaptation strategies they had developed to secure their livelihoods in the face of changes in the frequency, intensity, and duration of floods. These strategies included changing cultivation to flood- and drought-resistant crops, to crops that can be harvested before the flood season, or to varieties of rice that will grow high enough to remain above the water when the floods come (Mitchell, Tanner, and Lussier 2007). As one woman explained, "As we never know when the rain will come, we had to change. I started to change the way I prepare the seedbed so that we don't lose all our crops. I am also using different crops depending on the situation" (p. 6).

The women also were clear about what they need to better adapt to the floods: not only crop diversification and agricultural practices, but also skills and knowledge training to learn about flood- and drought-resistant crops and the proper use of manure, pesticides, and irrigation. These women have a strong understanding of the types of interventions required to ensure more sustainable agricultural processes in the face of these changes (Mitchell, Tanner, and Lussier 2007). These findings reaffirm a point made repeatedly in the literature on gender and the environment: women and men have distinct and valuable knowledge about how to adapt to the adverse impacts of environmental degradation (Gurung et al. 2006; Laudazi 2003; WEDO 2003).

More participatory research is needed into the adaptation strategies of women and men at the household and community levels in the face of existing climate change impacts on agricultural productivity and food security, including how these are manifested in different contexts. This research must ask the following questions: What aspects of women's and men's agricultural knowledge could contribute to effective adaptation? What are women and men already doing, and what do they identify as their needs and priorities? Future adaptation and agriculture policies should draw explicitly on these insights and seek to better support existing localized strategies.

The Importance of Voice

To ensure that climate change responses are effective—and uphold the principles of equitable and sustainable development—climate change policy and program design processes need to be gender sensitive; and they actively must seek and respond to the perspectives, priorities, and needs of *all* stakeholders, regardless of gender, age, or socioeconomic background (Polack 2008). Such efforts are key to ensuring that diverse perspectives are included and that valuable knowledge is not lost.

In the context of climate change, however, women and girls have remained conspicuously absent from decision-making processes at all levels. For example, at the 2007 13th Session of the Conference of the Parties (COP13; Bali, Indonesia, 2007), women made up only 28 percent of delegation parties and 12 percent of heads of delegations (Ulrike Röhr, personal communication). To challenge the unequal representation of women in climate change decision making and to promote a diversity of perspectives in international negotiations, GenderCC–Women for Climate Justice provided funding opportunities for women from the South to be included in the organization's delegation at the United Nations Framework Convention on Climate Change COP13. Each member of the delegation provided gender-focused input into conference debates; lobbied governments, constituencies, and networks to include gender considerations in the negotiations; and distributed position papers. Additionally, the GenderCC network planned daily women's meetings and held a capacity-building workshop for women (Röhr et al. 2008).

There also are numerous examples of cases in which women have not been consulted in the design and planning of community-level water or agricultural initiatives because doing so would require them to step outside their traditional, nonpublic roles into public and technical arenas for which they are seen as "unqualified" and "unsuited" (Fisher 2006). Lack of opportunity to feed knowledge into community- or national-level adaptation and mitigation strategies actually could jeopardize larger processes of reducing climate change and its impacts, and could undermine the effectiveness of projects at the local level. In the Kilombero district of Tanzania, for example, when a well built by a nongovernmental organization dried up shortly after it was created, it was revealed that the well's location had been decided by an all-male local committee. When development workers talked to the local women, they discovered that it is often the women's task to dig for water by hand, so they know the places that provide the best water yields. Since the incident, women have had more involvement in decisions about the location of wells (Fisher 2006). As well as helping to ensure that policy and programmatic responses to climate change take into account the diverse needs and priorities of both women and men, promoting women's and girls' meaningful participation in decision making can contribute to addressing gender inequalities by raising the profile and status of women and girls in the community and within society more broadly—helping to challenge traditional assumptions about their capabilities.

Critical Reflections

Clearly, there is much that can be learned from the existing body of knowledge and practice on gender and the environment; and it is essential that we build on this existing knowledge and experience rather than simply "reinventing the wheel." But there are significant weaknesses as well—notably, the lack of attention to intersecting forms of disadvantage, and the subsequent reliance on generalizations about women (and men) that cannot hold true for all people in all places. As emerges in much of the analysis above, the tendency within mainstream gender analysis of climate change is to conceptualize women everywhere as a homogenous, subjugated group— "the poorest of the poor"—irrespective of their location (rural or urban), their social classes or castes, their ages, their degrees of education or access to resources, their embeddedness in social networks, the numbers of dependents they support, and so forth. On the basis of this presumed universal vulnerability, the default has been to focus exclusively on women. To the degree that men are brought into view at all in gender analyses of climate change, it most often is to point the finger at "men in general" for causing climate change, with little qualification as to *which men* in particular we are talking about (for example, see Johnsson-Latham 2007).

Such representations are problematic on multiple fronts, particularly in their failure to account for the complex interactions between gender inequality and other forms of disadvantage and exclusion based on class, caste, region, age, race/ethnicity, and sexual orientation. Intersecting inequalities produce differing experiences of power and powerlessness between and among diverse groups of women and men, and these, in turn, enable or deny them certain choices—for example, determining whether migration in the face of environmental degradation is a viable option. (For the elderly or very young, those with limited resources as a result of economic marginalization or entrenched gender ideologies, or those facing cultural or religious restrictions on their mobility, such an option may not be feasible.)

Intersecting inequalities also mean that whereas men in most societies enjoy the benefits of male privilege, they may share with the women in their lives similar experiences of indignity, subordination, and insecurity as a result of discrimination or social and economic oppression (Esplen and Greig 2008). So rather than discounting men in gender analyses of climate change as if they are somehow nongendered and impervious to the harsh impacts of environmental degradation in contexts of economic or social marginalization, we need to find spaces within gender and climate change frameworks to acknowledge and communicate the vulnerabilities that some men also experience. Finding such space requires not only a better gender analysis, but also a stronger intersectional analysis that makes visible class oppression and economic marginalization alongside gender inequality as key axes of insecurity. It also requires that we move beyond framing the issues in terms of "vulnerable women," and focus instead on power relations within society—questioning who has the power to identify priorities and solutions and to shape debates and make decisions, and who does not.

It is in the light of our uneasiness with these broad generalizations about women and about men that we believe more context-specific participatory research is needed to illuminate the local realities, knowledge, and coping strategies of different groups of women and men in the face of accelerating climate change. By generating richer ethnographies of women's and men's lived realities, we can move toward more concrete, holistic understandings of how gender-related constraints—in combination with other intersecting forms of disadvantage—play out in particular contexts of impoverishment and environmental stress. Critically, we need to ensure that these local voices and perspectives reach national and international policy actors so that these crucial findings don't simply evaporate. Doing so will require investment in building the capacity of women and men to have the skills and confidence to engage in climate change debates at the local, national, regional, and international levels—for example, by supporting grassroots awareness-raising, confidence-building, and advocacy and leadership training programs. If we genuinely are to move toward more equitable, appropriate, and effective climate change policies and programs, this is perhaps the single most important step.

References

Aguilar, Lorena. n.d. "Climate Change and Disaster Mitigation: Gender Makes the Difference." International Union for Conservation of Nature, Gland, Switzerland.

APF/NEPAD (Africa Partnership Forum/New Partnership for Africa's Development Secretariat). 2007. "Gender and Economic Empowerment in Africa." Eighth Meeting of the Africa Partnership Forum, Berlin, Germany, May 22–23. http://www.africapartnershipforum.org/dataoecd/57/53/38666728.pdf.

Bartlett, Sheridan. 2008. "Climate Change and Urban Children: Impacts and Implications for Adaptation in Low- and Middle-Income Countries." Climate Change and Cities Discussion Paper 2, International Institute for Environment and Development, London. http://www.iied.org/pubs/display.php?o=10556IIED.

Brody, Alyson, Justina Demetriades, and Emily Esplen. 2008. "Gender and Climate Change: Mapping the Linkages—A Scoping Study on Knowledge and Gaps." BRIDGE–Institute of Development Studies, University of Sussex, Brighton, UK. http://www.bridge.ids.ac.uk/reports/Climate_Change_DFID_draft.pdf.

Chant, Sylvia. 2007. "Dangerous Equations? How Female-Headed Households Became the Poorest of the Poor: Causes, Consequences and Cautions." In *Feminisms in Development: Contradictions, Contestations and Challenges*, ed. Andrea Cornwall, Elizabeth Harrison, and Ann Whitehead, 35–47. London: Zed Books.

Esplen, Emily, and Alan Greig. 2008. "Politicising Masculinities: Beyond the Personal." Institute of Development Studies, University of Sussex, Brighton, U.K. http://www.siyanda.org/docs/esplen_greig_masculinities.pdf.

Fisher, Julie. 2006. "For Her, It's the Big Issue: Putting Women at the Centre of Water Supply, Sanitation and Hygiene. Evidence Report." Water Supply and Sanitation Collaborative Council, Geneva, Switzerland. http://www.genderandwater.org/page/5124.

Gurung, Jeannette D., Sheila Mwanundu, Annina Lubbock, Maria Hartl, and Ilaria Firmian. 2006. "Gender and Desertification: Expanding Roles for Women to Restore Drylands." International Fund for Agricultural Development, Rome. http://www.ifad.org/pub/gender/desert/gender_desert.pdf.

Hemmati, Minu. 2005. "Gender and Climate Change in the North: Issues, Entry Points and Strategies for the Post-2012 Process and Beyond." Genanet/Focal Point Gender Justice and Sustainability, Frankfurt, Germany.

Johnsson-Latham, Gerd. 2007. "A Study on Gender Equality as a Prerequisite for Sustainable Development: What We Know about the Extent to Which Women Globally Live in a More Sustainable Way Than Men, Leave a Smaller Ecological Footprint and Cause Less Climate Change." Environment Advisory Council, Ministry of the Environment, Stockholm, Sweden.

Kabeer, Naila. 2008. *Mainstreaming Gender in Social Protection for the Informal Economy*. New Gender Mainstreaming Series on Development Issues. London: Commonwealth Secretariat.

Lambrou, Yianna, and Grazia Piana. 2006. "Gender: The Missing Component of the Response to Climate Change." Food and Agriculture Organization of the United Nations, Rome. http://www.fao.org/sd/dim_pe1/docs/pe1_051001d1_en.pdf.

Laudazi, Marina. 2003. "Gender and Sustainable Development in Drylands: An Analysis of Field Experiences." Food and Agriculture Organization of the United Nations, Rome. http://www.fao.org/docrep/005/j0086e/j0086e00.htm.

Mitchell, Tom, Thomas M. Tanner, and Kattie Lussier. 2007. "We Know What We Need: South Asian Women Speak Out on Climate Change Adaptation." Action-Aid International and Institute of Development Studies, University of Sussex, Brighton, UK. http://www.eldis.org/go/display&type=Document&id=35433.

Moser, Annalise. 2007. "Gender and Indicators Overview Report." BRIDGE–Institute of Development Studies, University of Sussex, Brighton, U.K. http://www.bridge.ids.ac.uk/reports_gend_CEP.html.

Nair, Sumati, and Preeti Kirbat, with Sarah Sexton. 2004. "A Decade after Cairo: Women's Health in a Free Market Economy." *Corner House Briefing* 3. http://www.thecornerhouse.org.uk/item.shtml?x=62140.

Polack, Emily. 2008. "A Right to Adaptation: Securing the Participation of Marginalised Groups." *IDS Bulletin* 39 (4): 16–23.

Reuveny, Rafael. 2007. "Climate Change-Induced Migration and Violent Conflict." *Political Geography* 26 (6): 656–73.

Röhr, Ulrike. 2006. "Gender and Climate Change—A Forgotten Issue?" *Tiempo: Climate Change Newsletter* 59. http://www.tiempocyberclimate.org/newswatch/comment050711.htm.

———. 2008. "Gender Aspects of Climate-Induced Conflicts." *ECC-Newsletter* Special Edition, "Gender, Environment, Conflict."

Röhr, Ulrike, Gotelind Alber, Minu Hemmati, Titi Soentoro, and Anna Pinto. 2008. "Gender and Climate Change: Networking for Gender Equality in International Climate Change Negotiations—UNFCCC COP13/CMP3." http://www.gendercc.net/fileadmin/inhalte/Dokumente/UNFCCC_conferences/Report COP13_web.pdf.

Tanner, Thomas M., and Tom Mitchell. 2008. "Entrenchment or Enhancement: Could Climate Change Adaptation Help Reduce Chronic Poverty?" Working Paper 106, Chronic Poverty Research Centre, Manchester, U.K.

WEDO (Women's Environment and Development Organization). 2003. "Common Ground: Women's Access to Natural Resources and the United Nations Millennium Development Goals." WEDO, New York.

World Bank. 2007. *Global Monitoring Report 2007—Millennium Development Goals: Confronting the Challenges of Gender Equality and Fragile States.* Washington, DC: World Bank.

CHAPTER 6

The Role of Indigenous Knowledge in Crafting Adaptation and Mitigation Strategies for Climate Change in Latin America

Jakob Kronik and Dorte Verner

Indigenous peoples often are excluded or treated as secondary in the climate change debate. They often are considered simultaneously the most vulnerable and the most resourceful in adapting to climate change. Indigenous people we have interviewed in Latin America also perceive this apparent contradiction. In the Colombian Amazon, for example, it turns out that those of the indigenous peoples who have greater territorial autonomy— who derive their livelihood mostly from forest and water resources and maintain an active and engaged ritual life—see their livelihood most strongly affected. They place great value on gardens slashed in mature forest, planted with a high variety of species; depend heavily on fish and game for protein; and take care of their health with their own means and knowledge. Their livelihood rests on their ability to interpret regular natural cycles and act in accordance with them. Though they certainly have contact with mainstream society, are incorporated to some degree in the market economy, and have access to public health and education services, a large proportion of their livelihood depends on their knowledge, use, and management of forest and water resources. Our interviews showed these indigenous peoples to be most aware of and most vulnerable to the effects of climate change and variability (Kronik and Verner forthcoming).

Indigenous peoples' knowledge is treated primarily as historical and timeless data that can be merged into current plans and programs. However, this approach does not provide opportunities for understanding the specific political and historical circumstances that develop, maintain, and transmit indigenous knowledge. In this chapter, we suggest that the necessary conditions for developing indigenous peoples' knowledge and dialogue concerning the development of adaptation and mitigation instruments include political, social, and natural dimensions.

This chapter has three purposes: (1) to survey the social impacts of climate change on indigenous peoples in Latin America; (2) to explore how indigenous peoples have reacted to environmental change and, in this process how they have been shaping not only their societies, cultures, and capacity to adapt, but also their role in shaping nature; and (3) to address the role of indigenous peoples' knowledge in current and future climate change adaptation and mitigation efforts. Following the discussion of methodology, the chapter is organized in four sections. The first section presents the climate change projections for the Latin America and Caribbean region, with an account of where indigenous peoples live in relation to these projections, and a brief discussion of their vulnerability to climate change and the importance of understanding the indigenous context for the development of successful adaptation strategies. The second section presents the debates concerning the role of indigenous peoples' knowledge of natural resource management and the natural versus cultural shaping of landscapes and ecosystems. Each debate illustrates aspects of vulnerability, agency, and capacity to contribute to the adaptation strategies and mitigation instruments relevant to climate change. In the third section, we discuss the role of indigenous peoples' knowledge in crafting adaptation and mitigation strategies. Finally, in the fourth and concluding section, we discuss key questions on this basis, drawing on recent fieldwork in Latin America.

Delimitation and Methodology

This chapter provides an overview of climate change impacts on indigenous peoples and, in particular, the role of indigenous knowledge in crafting adaptation and mitigation strategies. Much of this overview draws on the hypotheses of Kronik and Verner (forthcoming) who argue that indigenous peoples in Latin America not only are particularly socially vulnerable to the impacts of climate change, but also play unique roles in climate change adaptation and mitigation because of their knowledge systems and rights. Although the hypotheses of Kronik and Verner generally hold true, certain caveats remain.

The lack of certainty results from the limitations of both the literature review and the undertaken fieldwork. A major issue for the literature review is that information pertaining to the diversity of indigenous culture, of ecosystems and natural resource management strategies, and of types of climate change impacts—floods, droughts, hurricanes, and the like—is limited.[1] Limitations arise from the relative newness of climate change as a field for analysis; and the fact that "indigenous" often is categorized under different terms, geographic regions, or ethnic groups.[2] Fieldwork methodology examining a diverse set of indigenous groups in Latin America was developed to better understand how these distinct and varied communities perceive the effects of climate change.[3]

The overall approach is based on the main types of climate changes besetting Latin America and the Caribbean (Verner forthcoming). The region is divided into three ecological areas: the Amazon, the Andes and sub-Andes, and the Caribbean and Mesoamerica. Using methods such as key informant interviews, participatory climate change impact scenarios, and institutional analyses, researchers studied inhabitants in these ecological regions in the countries of Bolivia, Colombia, Mexico, Nicaragua, and Peru. Kronik and Verner (forthcoming) provide an analysis of each of these ecological regions, as well as cross-regional comparisons.

Climate Change in Latin America and the Caribbean

Climate variations are driven partly by the uneven distribution of solar warmth; atmospheric moisture; and the interplay of individual and collective responses of the atmosphere, oceans, and land surfaces across Latin America (Christensen 2009). The variable impacts of climate change on different regions—many negative and some positive,—lead to less-robust projections than otherwise would be expected. Christensen's assessments, which support the findings of the latest Intergovernmental Panel on Climate Change report (Parry et al. 2007), show that Central and South America are very likely to warm during this century. Annual mean warming in southern South America is likely to be similar to global mean warming trends, whereas warming in the rest of the region will be more intense. This projection implies that temperatures in all seasons will continue to rise during the 21st century. Some aspects of temperature-related events (such as heat wave frequency and intensity) also are expected to change, unless the temperature increase is a result of entirely new circulation patterns. Likewise, increased warming trends lead to an enhanced risk of change to the seasonality of severe weather events, such as a longer hurricane season.

Precipitation predictions show differentiated impacts across the region. Annual precipitation is likely to decrease in most of Central America as the boreal spring becomes drier. Annual mean precipitation is projected to decrease across the Caribbean coasts of northern South America as well as over large parts of northern Brazil, Chile, and Patagonia. Simultaneously, precipitation is projected to increase in Colombia, Ecuador, Peru, around the equator, and in southeastern South America. The seasonal cycle modulates this mean change, especially over the Amazon basin where monsoon precipitation increases during the months of December to February and decreases during the months of June to August. Annual precipitation is likely to decrease in the southern Andes, especially during the summer season. As a caveat, changes in atmospheric circulation may induce high precipitation variability in local mountainous areas.[4]

Extreme events also are likely to change. Tropical cyclone activity in Central America and the Caribbean contributes to increased precipitation. Recent studies using improved global models suggest changes in the number and intensity of future tropical cyclones (hurricanes). A synthesis of the models' results indicates a warmer climate will increase the intensity of wind gusts and heavy precipitation during tropical cyclones, while it reduces the number of weak hurricanes.

Regions Inhabited by Indigenous Peoples

Currently, there are more than 600 different indigenous peoples in Latin America,[5] each with a distinct language and worldview (table 6.1). The majority of indigenous inhabitants live in the colder and temperate high Andes and Mesoamerica, located between southern Colombia and northern Chile and between southern Mexico and Guatemala. However, the majority of populations with distinct ethnolinguistic identities are found in the warm tropical lowlands, most notably in the Amazon rain forest. It is true that several indigenous cultures live in urban settings, but this chapter focuses on rural indigenous populations.

The populations can be described as living in areas either prone to rapid-onset hazards or prone to slow-onset hazards. In the Caribbean, the increased intensity of storms and hurricanes not only decreases access to natural resources (such as fisheries, forests, and arable land), but also

Table 6.1. Demographic Data on Indigenous Peoples in Latin America and the Caribbean

Country	Number of Peoples	Approximate number of indigenous individuals (thousands)	Percent of country population
Argentina[a]	31	650–1,100	2.5–5.0
Belize[a,b]	2	55	17.0–20.0
Bolivia[c]	36	5,200	62.0[h]
Brazil[a]	225[d]	700	0.4
Chile[c]	8	1,060	6.5
Colombia[c]	92[f]	1,400	3.4
Guatemala[a]	20[e]	6,000	60.0
Ecuador[c]	14	830	6.8
El Salvador[a,b]	3	400	8.0
Honduras[b,e]	7	75–500	1.0–7.0
Mexico[a,g]	61	12,400	13.0
Nicaragua[a]	7	300	—
Panama[a]	7	250	8.4
Paraguay	20	89	1.7
Peru	65	8,800	33.0
Suriname[a]	6	50	8.0
Uruguay	0	0	0.0
Venezuela[a]	40	570	2.2

Sources: Authors' elaboration, based on source material cited below in notes a–g. Layton and Patrinos (2006) is the source for information not accompanied by a note.

Note: — = not available.

a. Self-identification (IWGIA 2008).
b. Layton and Patrinos 2006.
c. Self-identification, VI Population Census, National Statistical Institute.
d. Instituto Socioambiental.
e. U.S. Central Intelligence Agency *World Fact Book.*
f. Indigenous peoples organizations.
g. Comisión de Desarrollo Indígena, Mexico.
h. Older than 15 years of age.

destroys infrastructure and personal belongings. An interesting finding in need of systematic inquiry is the apparent loss of land indirectly resulting from hurricanes. In Nicaragua, for instance, impenetrable forests that once prevented settlers and ranchers from accessing indigenous lands have been wiped out by hurricanes; and, as a result, there has been encroachment into indigenous lands (Kronik and Verner forthcoming). In the Andean region, water scarcity caused by the melting of glaciers and alterations in the hydrological regime is perceived as the main indirect impact of climate change and variability.

The elders and traditional leaders in the high Andes and the Amazon regions also are experiencing the social consequences of climate change.

Traditionally seen as local experts, these leaders lose credibility and well-being when climatic conditions become impossible to predict. Unprecedented changes in the timing of frost, harsh rain, and drought are disrupting the agricultural cycle in ways that no one ever would imagine. When such events recur, it undermines ritual practices, joint social memory, and the ability of elders to maintain social order. Such social upheaval, in turn, leads to serious consequences for local governance of natural resources. When traditional authorities cannot guarantee abundance and prosperity, their status falls and people look elsewhere for solutions to their problems, both by seeking other bodies of knowledge and by migrating.

In the Amazon, almost counterintuitively, the areas most affected by climate change and variability often are where indigenous peoples have greater territorial autonomy, derive their livelihood mostly from forest and water resources, and maintain an active and engaged ritual life. Such peoples place great value on the ecological diversity of mature forest. They depend heavily on fish and game for protein, and maintain health through their own means and knowledge. Their livelihoods rest on their abilities to interpret regular natural cycles and to act accordingly. Although they do have contact with mainstream society, participate moderately in the market economy, and have access to public health and education services, a large percentage of their livelihood depends on their traditional knowledge and management of natural resources. However, this knowledge is becoming obsolete under the influence of climate change, and their daily practices increasingly are failing to respond effectively to changes in precipitation patterns. Based on interviews, these indigenous peoples emerge as those people most aware of and most vulnerable to climate change and its effects.

By contrast, indigenous peoples living close to urban centers in the Andes region, the Amazon, and in parts of Guatemala, Honduras, and Nicaragua depend on secondary forest horticulture, cash crops from alluvial soils, commercial fishing, wage labor, tourism, and sale of handicrafts for their livelihood. Their social systems are affected by climate change to the extent that they use river and forest resources. However, these indigenous peoples are less in tune with the seasonal calendar, their traditional knowledge is more limited, and ritual specialists generally play a smaller role in their lives. As a result of their integration into mainstream culture, they enjoy greater access to markets and public health and education services—all of which provide a buffer against many effects of climate change—than do groups living more traditional lifestyles (Kronik and Verner forthcoming).

Nature of Threats and Adaptation Strategies

Across the Latin American continent, there are many threats common to most indigenous peoples. A recurrent lament voiced during all the field visits was that the cultural adaptation strategies developed to tackle the "normal span of variation" no longer are sufficient, given the increased unpredictability in seasonal variation. The disturbance of the annual ecological calendar and the resulting disruption of the agricultural calendar probably are the most serious threats of all. The annual succession of seasons is of utmost importance for indigenous peoples. This rhythm orders the timing of the horticultural cycle and the ritual practices that help prevent illnesses and promote human well-being, and is crucial for the reproduction of wildlife. Unpredictability of the seasonal variation causes food insecurity and undermines the array of solutions provided by cultural institutions and authorities.

Changes in the timing of the dry and rainy seasons, alterations in the flood pulses of the rivers, changes in winds, and abnormal cold and heat episodes have become apparent during the last decade. Indigenous peoples are keen observers of natural rhythms, and they have accumulated a large and sophisticated body of knowledge about annual seasonal cycles. Most indigenous peoples in Latin America are rain-fed agriculturalists, producing crops such as maize or potatoes. Other small groups of indigenous people include horticulturalists, fishers, hunters, and those who combine these practices with rain-fed agriculture. These livelihood systems are tied closely to predictable and well-established seasons. Indigenous peoples possess a strong awareness of complex ecological indexes of the timing of seasons—seasons that were clearly established and well known to them until a decade ago. These natural rhythms serve to regulate, defend, and maintain life by governing the interrelation of water, wind, heat, fish, terrestrial fauna, insects, wild fruits, and human activities. According to key informants from across Latin America, the natural signs and indicators they now perceive are alarming. Seasons have become irregular; and the once-regular flow and descent of rivers is now out of synchronicity with seasonal events (such as the fall of wild fruits) that directly affect livelihoods.

Indigenous Peoples' Knowledge

The debates regarding indigenous peoples' knowledge and the role it plays with respect to natural resource management in general and biodiversity in particular have increased over the last four decades, achieving international recognition in the Convention on Biological Diversity (article 8j), in International Labour Organization Convention 169,[6] and with the United Nations Declaration on the Rights of Indigenous Peoples. Though the Declaration was not a binding document, its adoption was seen as a huge step forward in the recognition of the rights of indigenous peoples.

In the literature, indigenous peoples' knowledge[7] most often has been defined in juxtaposition to scientific knowledge, modern knowledge, and western knowledge. On that basis, indigenous knowledge may appear to scientists as "myth"—vague, subjective, context-dependent, open to multiple interpretations, and embedded in cultural institutions such as kinship. Cruikshank (2001) argues that local people may characterize scientific knowledge in similar terms: as "illusory, vague, subjective, context-dependent and open to multiple interpretations and embedded in social institutions like distant universities" (p. 390).

An area of importance is the recent debate on the role of indigenous peoples' knowledge in responding to environmental change. This debate includes (1) the nature of indigenous knowledge versus scientific or modern knowledge production, and (2) culture versus nature. Since the early 1980s, there has been an increasingly intense debate on the forces and drivers behind the shaping of landscapes, the ecological character of ecosystems such as rainforests, and thus the extent of indigenous peoples' agency and historical capacity to respond to environmental change. A branch of this debate discusses how important biological resources are preserved in the absence or presence of human activity.

Role of Indigenous Peoples' Knowledge and Its Articulation

Indigenous peoples' knowledge and its articulation, combined with other forms of knowledge (for example, modern scientific knowledge), can play a role in crafting adaptation responses. Discussing the Achuar of the Ecuadorian Amazon, Descola (1994) presents the debate addressing the articulation of knowledge systems. He shows how the quality and diversity of the material outcomes of activities like gardening depend on the ability to draw on insights developed through cultural logic, rather than on what he calls "the sphere of practical reason" (such as environmental constraints and access to available workforce). A further understanding of this cultural logic requires a closer look at the production and reproduction of knowledge, eliciting the relevant learning processes and exploring the mutually related practices and institutions (Kronik 2001).

The idea that objective and universal knowledge is free of the influence of special interests is not in conflict with the positivist notion of one objective truth. However, other lines of thought must be consulted to understand how different individuals can view the same phenomena differently. Unlike positivism, social constructionism emphasizes the negotiation of processes leading to the establishment of knowledge, claiming that all knowledge and facts are products of complex processes of inquiry, negotiation, and institutionalization (Latour 1999). Positivism and essentialism, with their faith in absolute objective truth, cannot explain this. Within the framework of social constructionism, facts are seen as constructs developed out of culturally and historically specific contexts (McCarthy 1996; Burr 1995). In other words, facts are dealt with as fabricated knowledge. So, whenever one refers to knowledge as information or fact, it comes with the baggage of social and cultural history.

Beginning in the early 1980s, a group of academics and practitioners embarked, implicitly or explicitly, on a political project to counter the centralized, technically oriented, (mega-)project solutions to development, addressing in part the failure of the "green revolution" technologies to reach the poorest farmers (Agrawal 1995; Chambers and Jiggins 1986). One of the most influential local knowledge proponents, Chambers (1980), emphasizes inequality, power, and sense of prestige as some of the differences in relation to modern scientific knowledge. He claims that scientific knowledge "is centralized and associated with the machinery of the state; and those who are its bearers believe in its superiority" (p. 2). Indigenous technical knowledge, in contrast, is scattered and associated with low-prestige rural life. Even those who are its bearers may believe it to be inferior (Agrawal 1995; Chambers 1980). This was part of a political project to establish a dichotomy between rural people's knowledge and western, scientific knowledge of modernity. Separately as well as together, the academics and practitioners involved conceptualized what they saw as a body of knowledge that had as its main common denominator its difference from scientific knowledge. The concept of local knowledge springs from a process of politicization that has shaped the ways in which it is understood today. As argued by Bebbington (1994), the first sense in which local/indigenous knowledge is constructed is as a concept within an academic community. Simply by naming it, this community of scholars and practitioners[8] has created the idea that such a coherent body of knowledge exists.

Another epistemological paradigm maintains that local knowledge is based on whether a knowledge system can be considered closed or open. Within this paradigm, cultural flows bring in new insights not tied to a specific location or context. These insights challenge established explanations, arguing that knowledge systems are context-driven rather than adhering to any sense of universal truth. Hence, all knowledge production will depend on contextual conditions and relations. Descola (1994), Giddens (1979), Sahlins (1985), and others more recently have demonstrated that there is no such thing as a closed knowledge system. All types of knowledge constitute both empirical ideas and the principles that underlie their formation, organization, and meaning. All types of knowledge production are influenced by many sources, in varying forms and to varying degrees, including modern ideas and technologies. Giddens argues that the imagined closed character of a system is opened when people, as part of their practices, interpret concepts into a different context than the system.

Is local knowledge produced by different processes than scientific knowledge? Bell (1979) and Howes and Chambers (1979) offer two contrasting views. Howes and Chambers, seemingly inspired by Lévi-Strauss (1962), present the notion of *bricolage*, which proposes that local and scientific knowledge production differs with respect to analytical capacity. In their words, "An important difference between science and ITK [indigenous technical knowledge] lies in the way which phenomena are observed and ordered. The scientific mode of thought is characterized by a greater ability to break down data presented to the senses and to reassemble it in different ways" (p. 6). Additionally, a parallel body of literature critically examines the nature of knowledge production. Authors like Haraway (1988), Knorr-Cetina (1995), and Latour and Woolgar (1986) have documented how different representations of scientific production are inherently related to the contextual conditions under which they are made. These findings contribute to an understanding of scientific knowledge production as being developed and maintained within so-called cultures of science. Such cultures are composed of, among other things, incentive structures, specific environments, and varying types and degrees of influence and demands that certain clients exert on knowledge production. Authors like Harding (1994) and Horton (1967) call into question the claims of universality that often are made for science. Historically, there are good reasons to believe the research activities of Lévi-Strauss and Bell have been formalized into institutions, specializations, and complex learning systems. Lévi-Strauss (1962) and Rhoades and Nazarea (1999) speak against the earlier and more common notion that traditional knowledge is generated either from a conscious pursuit arising from a pressing "need to know" or from "accidental circumstances." Lévi-Strauss argues that indigenous peoples not only have developed impressive folk taxonomies and invaluable technologies, such as pottery making, weaving, and agriculture; but also have demonstrated highly developed knowledge processes and social mechanisms in developing these taxonomies and technologies.

Knowledge Institutions—Developers and Stewards of Knowledge and Reflexive Capacity

Basic to many Amerindian[9] peoples' understanding of the relationship between society and nature is the notion of maintaining balance. Thus, many indigenous peoples believe that human, natural, and cosmological realms are linked; and that these realms need to be in balance or equilibrium. These balances constantly are in flux, and living and acting involve negotiating them. So when changes occur—for example in climatic conditions—people look to themselves, their social institutions, and their rituals for the causes of such imbalances. For instance, they may try to discover if there is something wrong with the way they conduct their own lives and may try to amend them. If this is not possible, they seek other means to restore the balance between the social and the natural. Although these cultural practices vary from place to place, they share a striving for balance, based on trusted social and cultural knowledge and practices.[10] Indigenous peoples' knowledge systems are based on experimenting with nature, and they contain a stock of knowledge developed over time and passed on through generations. The ability to predict and interpret natural phenomena such as climatic conditions not only has been vital for survival and well-being; but also has been instrumental in the development of social structures, trust, and authority. In summary, there is a mutual influence between the societal knowledge generation about nature's cycles and certain cultural practices leading to the creation of cultural capital—which, in turn, is reproduced through practices and rituals. Cultural institutions are developed around these regularized practices and rituals; and they serve to maintain, develop, and dispute information and thereby contribute to the social generation of knowledge.

Social order and institutions exist in the minds of community members and continuously are produced through ongoing activities (Garfinkel 1967). The "glue" of institutions is the shared meanings developed, maintained, and sanctioned among their members and affected outsiders. The institution thus is constructed and reproduced by its members, and depends on their continued participation. People's participation is subject to competing offers and constraints within the social context of which it is part. The resilience of an institution, therefore, is subject to its continued relevance for its members. However, institutions like knowledge do not necessarily change overnight. They, too, have a built-in, self-perpetuating dynamic shaped by routine activities and established, shared meanings. This inertia of institutions has a structural influence on people's activities (Giddens 1984). From the common-sense idea that institutions and the people forming them have an interest in developing their knowledge of how to tackle a changing context, it follows that they also have an interest in strengthening their aggregate reflexive capacity. This strengthening may be achieved through strengthening the basic and specialized learning processes, by providing access to and encouraging participation in relevant institutions and practices, and by removing constraints to creative activities.

How can knowledge systems play a role in crafting adaptation responses? Rather than the modernist distinctions between "natural" and "social" knowledge, future efforts to strengthen adaptation to and mitigation of climate change may benefit from shifting the axis to emerging intersections between global and local knowledge. Comparative approaches can examine how local values contribute perspectives essential for the broader understandings now required to address global issues, such as how to deal with climate change. As Cruikshank (2001) emphasizes, both historical and comparative approaches take account of the power dynamics underlying any production of knowledge. At the same time, those approaches insist that local knowledge be taken seriously and given opportunities to interrogate scientific perspectives, a process already occurring in a variety of settings.[11]

The current development of adaptation strategies and mitigation instruments may improve from accepting and institutionalizing a dialogue between global and indigenous peoples' knowledge systems—not just to enrich them with context-dependent data, but also to see that the strategies and instruments reach their objectives in the social context where they are targeted. The local and global perspectives contribute values, norms, and rules.

All over the Americas, there are accounts of Amerindian peoples referring to the importance of human agency, choice, responsibility, and the consequences of human behavior (Kronik and Verner forthcoming; Cruikshank 2001).

Main Challenges in Crafting Adaptation Responses from a Dialogue between Knowledge Systems

A conscious approach to adaptation and mitigation includes an understanding of how knowledge systems developed under different circumstances and with different drivers and processes will influence social institutions. However, some indigenous peoples are compelled to change their livelihoods so dramatically that they lose vital conditions for the development and reproduction of their culture. Indigenous knowledge, institutions, and practices may be rendered superfluous, lost, or temporarily forgotten. There is an urgent need to support and develop monitoring of the changing climate. This effort includes larger monitoring systems that provide reliable, responsive, and relevant local information, as well as local monitoring and forecast systems. These systems should be combined with access to knowledge on how to interpret these changes and impacts to local livelihood strategies; and they raise important issues about information taxonomy, communication methods, and how to deal with uncertain data. Some of the experiences are referred to in the Intergovernmental Panel on Climate Change report (Parry et al. 2007, pp. 591–605).

Indigenous peoples' knowledge is treated primarily as historical and timeless data that can be merged into current plans and programs. However, this approach does not provide opportunities for understanding the specific political and historical circumstances that develop, maintain, and transmit indigenous knowledge.

Necessary conditions for the development of indigenous peoples' knowledge and dialogue concerning the development of adaptation and mitigation instruments include political, social, and natural dimensions. To craft adaptation responses to local and global knowledge systems, negotiation, design, and implementation must be participatory so that relevant processes, institutions, and practices are protected, consulted, and included.

Recent literature argues that the ability to withstand shocks and stresses to livelihoods is especially important in adapting to climate change and variability, and thereby is linked to vulnerability (Thomas et al. 2007). The World Bank (2000) links such factors as reliance on natural resources to high levels of vulnerability and low adaptive capacity in the developing world. However, as mentioned by Thomas et al. (2007) citing Salinger, Sivakumar, and Motha (2005), the resilience of human societies may be enhanced positively by people countering vulnerability, if wider dimensions of livelihood change permit this to occur (Robledo, Fischler, and Patiño

2004)'. These wider dimensions include well-functioning sets of regularized practices. Such institutions have been developed and actively maintained in indigenous communities; and they are a key identifier of the well-being of indigenous communities and their responsive capacity to social changes.

Natural Versus Cultural Dimensions

Until recently, research into the environmental history of Latin America has been influenced by antiquated Eurocentric myths (see Bowden 1992, and Turner and Butzer 1992). Endfield and O'Hara (1999b) argue that these myths have contributed to two allied explanations of environmental degradation. The first explanation is suggestive of a pristine environment on the eve of conquest, with the inference that indigenous populations lived in harmony with the environment and refrained from altering the landscape, which then was devastated by European colonization and their land-use systems (Denevan 1992). The second explanation acknowledges the negative impacts wielded by precontact land-use systems, but attributes accelerated landscape degradation to the European introduction of plow agriculture and livestock (Shelter 1991; Sale 1990). The debate still is contentious with respect to the effect of indigenous peoples' management of natural resources and landscapes. Whereas Betty J. Meggers (cited as a personal comment in Mann 2005, p. 4) and Dean Snow (1995) dispute the evidence that available data are misinterpreted based on perverse political correctness, and that the ethnohistorical record can be interpreted to support any agenda; other authors claim that indigenous populations all over Latin America had severe effects on their landscapes. Researchers show that, in the Amazon, indigenous peoples developed more than 12 percent of the entire region into anthropogenic or cultural forests, thus changing the ecological composition (Posey and Balée 1989, p. 2). For instance, the flood plains of Bolivia (first photographed by Denevan in the 1960s) had been shaped by a prosperous society, one whose existence had been forgotten (Mann 2005; Balée and Erickson 2005). Preconquest landscapes also have been found modified and severely degraded in many locations, with evidence of accelerated anthropogenic erosion spanning at least 3,600 years (Endfield and O'Hara 1999a, p. 404; Frederick 1995; O'Hara, Metcalfe, and Street-Perrott 1994; O'Hara, Street-Perrott, and Burt 1993). Migration, death, the spread of diseases, and oppression, combined with new tenure systems and land-use practices from the conquest onward, had devastating effects on indigenous populations and resource management. In Michoacán, Mexico, the environment was stressed further by a combination of increasingly intense droughts (O'Hara, Metcalfe, and Street-Perrott 1994), resource monopolization by emerging haciendas (Endfield 1997), and the demographic pressure of a burgeoning mestizo population (Butzer 1992). What we can learn from this is that indigenous peoples have contributed to the shaping of the Latin American landscapes and ecosystems.

The term *biodiversity* generates images of peaceful landscapes, leisure time, exotic animals, and beautiful plants. For some people, biodiversity raises concerns about the loss of valuable resources; and it does not present itself as the highly contested question it actually is. Actors interested in using, conserving, and benefiting economically from biodiversity are involved directly or indirectly in disagreements about the rights to access and ownership of biodiversity and associated knowledge. The conflicts over rights to knowledge and rights to plant genetic material certainly are among the most heated issues; and they provide insight into the current debate regarding the objectives and means of mitigation instruments, such as reducing emissions from deforestation and forest degradation (REDD) and biofuels. During the 1990s, indigenous peoples' organizations emphasized this debate. During this period, the apparently irreconcilable gap between indigenous peoples' understanding of common property rights to biodiversity and associated knowledge, on the one hand, and the biotechnological industry's demands for universal intellectual property rights based on patents, on the other, had become a point on the international agenda. Until the late 1970s, both the International Union for Conservation of Nature and the World Wildlife Fund favored establishing national parks as a key strategy to protect wildlife and their habitats from so-called destructive human activities (Myers 1984). The "fencing, policing and management strategy" has created a series of conflicts with the original users of these areas and among different interest groups within the conservation movement. The internal dispute largely was between actors with predominantly aesthetic motivations and those with a predominantly "natural-historical" motivation.

New Interests and New Players on the Scene

During the late 1980s, utilitarian and ethical motives for the conservation of biodiversity were being merged. Conservationist and broader environmental or developmental institutions began to develop common objectives, policies, strategies, and concrete actions. Other actors (such as indigenous peoples' organizations, industry, and academics) published their concerns in response to these efforts. They raised issues about intellectual and other kinds of property rights, benefit sharing, cultural diversity, and local knowledge. The structure of the international arena was changing, and new legal instruments began to emerge. One of the major results of this process is the new space for negotiation where adversaries can share at least one goal—a successful resolution of the conflict (Fowler 1994). However, to get this far—in fact, to enter into such negotiations constructively or at all—relevant stakeholders must realize that they have something to gain from such negotiations.

Adaptation and Mitigation

This section explores the role of indigenous peoples in the context of adaptation to and mitigation of climate change.

Adaptation

The notion of adaptation has been well debated in relation to indigenous peoples. Before it was linked to human systems by Steward (1937), it could be traced back to evolutionary biology, which refers to an organism or system's ability to cope with environmental stress in order to survive and reproduce (Futuyma 1979; Smit and Wandel 2006). Steward uses "cultural adaptation" to describe the adjustment of societies to the natural environment through subsistence activities (Butzer 1989). Denevan (1983) defines cultural adaptation as a "process of change in response to a change in internal stimuli, such as demography, economics and organization" (p. 401), thereby broadening the range of stresses to which human systems adapt beyond just the biophysical (Smit and Wandel 2006). Denevan (1983) further considers cultural practices, which enable human societies to develop and survive, as adaptations and believes these can be recognized analytically on the bases of behavior and innovation. Butzer (1980) was among the very first to link climate change to adaptation, connecting human ingenuity and long-term planning with predictions of climate change effects on world food supplies.

Mitigation

In Latin America, many rural populations are affected by climate change and variability (Kronik and Verner forthcoming). Currently, mitigation efforts from the local to the global level are being developed, and modalities are being negotiated. Instruments such as hydropower dams, biofuels, and forest protection schemes—originally developed to meet other objectives—are being rediscovered within the climate change debate as both panacea and reason for concern. In addition to biofuels, one of the more discussed instruments is avoided deforestation, REDD. For Latin America and the Caribbean, where the burning of forests is one of the main sources of greenhouse-gas emissions, the rights of indigenous peoples and their role in protecting forest areas are particularly relevant. Recent studies show that areas governed by indigenous peoples are less prone to deforestation. This growing international recognition was fought for by indigenous peoples' organizations and advocates, who claim indigenous peoples' rights to land and resources based on the beneficial effect of indigenous practices in support of biodiversity. Nepstad et al. (2006) shows that the inhibitory effect of indigenous lands on deforestation was strong after centuries of contact with national society. Indigenous lands occupy one fifth of the Brazilian Amazon—five times the area under protection in parks— and currently are the most important barrier to Amazon deforestation (Nepstad et al. 2006, p. 65). As the protected-area network expands in the coming years to include 36–41 percent of the Brazilian Amazon, the greatest challenge will be successful reserve implementation in areas at high risk of frontier expansion. This success will depend on broadly based political support.[12] Therefore, it is important to be thinking in terms of indigenous peoples' rights when designing and negotiating the new mitigation instruments. Lessons learned from the biodiversity and protected-area debate of the past two decades bring useful insights to the shaping of these instruments. Support for indigenous rights may help governments achieve carbon emission reduction targets.

In indigenous communities, cultural institutions play a significant role in the adaptation to climate change and variability. As regularized practices, these institutions are important in shaping natural resource management (World Bank 2002, 2006). In addition, cultural institutions have the potential to play an important role in the context of mitigation instruments, such as the implementation of forest protection policies. So far, only limited knowledge is available on the role of cultural institutions in the context of climate change and variability. Indigenous peoples have an important role in adaptation—by promoting the incorporation of experience with new and previously unknown phenomena (changing climate, and thus flora and fauna)—and perhaps in the context of mitigation. Several Latin American and Caribbean countries currently are launching programs aimed at generating direct or indirect benefits in terms of REDD.

Conclusion

The research conducted by Kronik and Verner (forthcoming) brings out five important messages. First, adaptation and mitigation will fall into a certain context. This context includes social, economic, and natural forces as well as actors. Second, social, political, cultural, and environmental forces determine vulnerability in terms of exposure and sensitivity; and they shape local adaptive capacity. Third, larger processes and instruments such as rights, laws, and economic drivers are at least as important to address as are local-level adaptation and mitigation. Fourth, no adaptation is started for climate change purposes alone. It is imperative to incorporate adaptation into overall development policies, plans, and programs. Fifth, cultural values that deal with uncertainty and variability, developed within indigenous peoples' institutions over time, may be the main contribution in developing adaptation and mitigation strategies.

It has become apparent that indigenous peoples' knowledge is vast with respect to natural resource management in Latin America; that it continually is being produced and maintained through social processes; and that these processes depend on social, cultural, and biophysical conditions, as well as on access to certain natural resources. When these conditions change, knowledge production also may change. Drawing on the findings from Kronik and Verner (forthcoming) and the above analyses, we can see that the capacities to adapt to climate change are under stress in the major ecogeographical regions of Latin America and the Caribbean. However, we also may see that indigenous peoples' knowledge plays a significant role in the protection of large ecosystems, and that it will be difficult to achieve climate change adaptation and mitigation without taking action to strengthen the necessary conditions for continued use and development of indigenous knowledge.

Adaptive capacity is based on available assets and conditions such as knowledge, impact intensity, and the level of vulnerability. These assets and conditions will vary among different indigenous peoples, depending on cultural features, social capital, productive practices, socioeconomic contexts, and political situations. Thus, indigenous peoples have to adjust their livelihood strategies to an altered environment, the disruption of the agricultural calendar, droughts, heavy rains, and the changing of water flows. In some cases, they cannot find answers to their needs in their toolbox of cultural knowledge and adaptation strategies—a problem that is compounded by the lack of state institutions and that ultimately leads to increased migration and social change. The combined pressure of climate change–induced decreased access to a resource, the institutional vacuum, the loss of esteem for authorities, and the loss of trust in cultural knowledge adds threats to traditional culture and religion and accentuates the need to address governance. Moreover, access to information, mutual understanding of the role of knowledge systems for the interpretation of the phenomena and effects of climate change and variability, and access to resources and relevant institutional capacities need to be improved.

Indigenous peoples' knowledge also has an important positive role in the context of adaptation. The current processes, through which traditional knowledge is lost on the current generation of farmers, jeopardize adaptation to the climate changes. For example, today there are few farmers who remember and use indicators such as the behavior of birds and insects—indicators that guided their predecessors in the agricultural calendar. Certain adaptation measures actually may turn out to be maladaptation. Take the case of expansion into new crops, such as irrigated onion, which will affect the use of other natural resources, such as soil and water. Combining indigenous peoples' knowledge, both in terms of agricultural techniques and the interpretation of biological indicators of change, with other local and government-supported adaptation schemes will enhance the effort's probabilities of success and cost effectiveness.

In the Latin American and Caribbean region, it is difficult to imagine that there can be much success with REDD without indigenous peoples' participation simply because they control (and often own) large tracts of dense forest. During recent years, indigenous peoples' organizations and networks have expressed serious concern over having their autonomy and authority undermined by entering into government-negotiated REDD agreements. It is necessary to find ways of recognizing and entering into agreements with indigenous peoples in areas such as forest protection, if they are to become an effective part of climate change mitigation mechanisms in the region.

Notes

1. As an example of the literature limitations, consider the bibliographical database ISI Web of Knowledge, and our experience with it. The database comprises the Social Science Citation Index (1956 forward), the Science Citation Index Expanded (1900 forward), and the Arts and Humanities Citation Index (1975 forward), as well as the huge collection of environment and development

literature of the Danish Centre of International Studies and Human Rights. Nonetheless, database searches including the terms "indigenous," "climate change," and "social impact" gave zero hits.

2. Relevant combinations of search terms—such as social impact/livelihood/ institution*/health, with indigenous/tribal/native/cultur*/, and climate change/ drought/season*/flood/precipitation/glaciers/hurricane*—have been made to identify and substantiate principal sources and debates (supplemented with the "snowball method") with the risk of focusing more on certain research areas than others.

3. The fieldwork strategy is further elaborated in Kronik and Verner (forthcoming).

4. It is uncertain how annual and seasonal mean rainfall will change over northern South America, including the Amazon forest. In some regions, there is qualitative consistency among the simulations (rainfall increasing in Ecuador and northern Peru, and decreasing at the northern tip of the continent and in the southern portion of northeast Brazil).

5. Indigenous peoples are an important and diverse segment of many Latin American countries. There is some debate about how many indigenous individuals there are in Latin America. National statistics estimate around 28 million; other estimates range from 34 to 43 million, reflecting the use of various definitions and calculation methods (see Layton and Patrinos 2006). The public policy ramifications of estimating population size are widely recognized as influencing the degree to which access to services, resources, and rights is given or limited. In table 6.1, the best available estimates have been gathered from various sources to provide an overview of the number of indigenous peoples, the number of indigenous individuals, and the relative sizes of the indigenous and national populations.

6. Convention 169 concerns indigenous and tribal peoples in independent countries.

7. This knowledge is variously referred to as indigenous knowledge, local knowledge, indigenous technical knowledge, or traditional ecological knowledge.

8. The community of scholars and practitioners is identified by the following works: Altieri (1987); Chambers, Pacey, and Thrupp (1989); Gliessman (1990); Warren, Slikkerveer, and Brokensha (1991); Warren, Slikkerveer, and Titiola (1989).

9. "Amerindian" refers to the indigenous peoples of the Americas.

10. For further analyses of traditional climate knowledge in the Andes, see Orlove, Chiang, and Cane (2000).

11. See also Agrawal (1995), Cronon (1992), Cruikshank (2001), Kearney (1994), Sillitoe (1998), Watson-Verran and Turnbull (1995), and Usher (2000).

12. "From satellite-based maps of land cover and fire occurrence in the Brazilian Amazon, the performance of large uninhabited parks (> 10.000 ha) and inhabited (indigenous lands, extractive reserves, and national forests) reserves were compared . . . No strong difference was found between parks and indigenous peoples' lands. However, uninhabited reserves tended to be located away from areas of high deforestation and burning rates, while in contrast, indigenous lands were often created in response to frontier expansion, and many prevented deforestation completely, despite high rates of deforestation along their boundaries" (Nepstad et al. 2006, p. 65).

References

Agrawal, Arun. 1995. "Dismantling the Divide between Indigenous and Scientific Knowledge." *Development and Change* 26 (3): 413–39.

Altieri, Miguel A. 1987. *Agroecology: The Scientific Basis of Alternative* Agriculture. Boulder, CO: Westview Press.

Balée, William. 1989. "The Culture of Amazonian Forests." In *Resource Management in Amazonia: Indigenous and Folk Strategies,* Advances in Economic Botany, vol. 7, ed. Darrell Addison Posey and William Balée, 1–21. New York: New York Botanical Garden Press.

Balée, William, and Clark L. Erickson, eds. 2005. *Time and Complexity in Historical Ecology: Studies in the Neotropical Lowlands.* New York: Columbia University Press.

Bebbington, Anthony J. 1994. "Composing Rural Livelihoods: From Farming Systems to Food Systems." In *Beyond Farmer First: Rural People's Knowledge, Agricultural Research and Extension Practice,* ed. Ian Scoones and John Thompson, 88–96. London: Intermediate Technology Publications.

Bell, Martin. 1979. "The Exploitation of Indigenous Knowledge or the Indigenous Exploitation of Knowledge: Whose Use of What for What?" *IDS Bulletin* 10 (2): 44–50.

Bowden, Michael Joseph. 1992. "The Invention of the American Tradition." *Journal of Historical Geography* 18: 3–26.

Burr, Vivien. 1995. *An Introduction to Social Constructionism.* London: Routledge.

Butzer, Karl W. 1980. "Adaptation to Global Environmental Change." *Professional Geographer* 32 (3): 269–78.

———. 1989. "Cultural Ecology." In *Geography in America at the Dawn of the 21st Century,* ed. Gary L. Gaile and Cort J. Willmott, 97–112. Columbus, OH: Merrill Publishing.

———. 1992. "The Americas Before and After 1492: An Introduction to Current Geography Research." *Annals of the Association of American Geographers* 82 (3): 345–68.

Chambers, Robert. 1980. "Understanding Professionals: Small Farmers and Scientists." IADS Occasional Paper, New York, NY: International Agricultural Development Service.

Chambers, Robert, and Janice Jiggins. 1986. *Agricultural Research for Resource Poor Farmers: A Parsimonious Paradigm.* IDS Discussion Paper 220. Brighton, U.K.: Institute of Development Studies.

Chambers, Robert, Arnold Pacey, and Lori Ann Thrupp, eds. 1989. *Farmer First: Farmer Innovation and Agricultural Research.* London: Intermediate Technology Publications.

Christensen, Jens Hesselbjerg. 2009. "Climate Change and Climatic Variability." Unpublished manuscript, World Bank, Washington, DC.

Cronon, William. 1992. "A Place for Stories: Nature, History, and Narrative." *Journal of American History* 78 (4): 1347–76.

Cruikshank, Julie. 2001. "Glaciers and Climate Change: Perspectives from Oral Tradition." *Arctic* 54 (4): 377–93.

Denevan, William M. 1983. "Adaptation, Variation and Cultural Geography." *Professional Geographer* 35 (4): 399–407.

———. 1992. "The Pristine Myth: The Landscape of the Americas in 1492." *Annals of the Association of American Geographers* 82 (3): 369–85.

Descola, Philippe. 1994. *In the Society of Nature: A Native Ecology in Amazonia.* Cambridge, U.K.: Cambridge University Press.

Endfield, Georgina H. 1997. "Myth, Manipulation and Myopia in the Study of Mediterranean Soil Erosion." In *Archaeological Sciences 1995,* ed. Anthony Sinclair, Elizabeth A. Slater, and John Gowlett, 241–48. Oxford, U.K.: Oxbow.

Endfield, Georgina H., and Sarah L. O'Hara. 1999a. "Degradation, Drought, and Dissent: An Environmental History of Colonial Michoacán, West Central Mexico." *Annals of the Association of American Geographers* 89 (3): 402–19.

———. 1999b. "Perception or Deception? Land Degradation in Post-Conquest Michoacán, West Central Mexico." *Land Degradation and Development* 10: 383–98.

Fowler, Cary. 1994. *Unnatural Selection—Technology, Politics, and Plant Evolution.* Yverdon, Switzerland: Gordon and Breach Science Publishers.

Frederick, Charles. 1995. *Fluvial Responses to Late Quaternary Climate Change and Land Use in Central Mexico.* Austin, TX: University of Texas.

Futuyma, Douglas J. 1979. *Evolutionary Biology.* Sunderland, MA: Sinauer Associates.

Garfinkel, Harold. 1967. *Studies in Ethnomethodology.* Englewood Cliffs, NJ: Prentice-Hall.

Giddens, Anthony. 1979. *Central Problems in Social Theory: Action, Structure and Contradiction in Social Analysis*. Berkeley, CA: University of California Press.

———. 1984. *The Constitution of Society: Outline of the Theory of Structuration*. Berkeley, CA: University of California Press.

Gliessman, Stephen R, ed. 1990. *Agroecology: Researching the Ecological Basis for Sustainable Agriculture*. New York: Springer-Verlag.

Haraway, Donna. 1988. "Situated Knowledges: The Science Question in Feminism as a Site of Discourse on the Privilege of Partial Perspective." *Feminist Studies* 14 (3): 575–99.

Harding, Sandra G. 1994. "Is Science Multicultural? Challenges, Resources, Opportunities, Uncertainties." *Configurations* 2 (2): 301–30.

Horton, Robin. 1967. "African Traditional Thought and Western Science, Parts 1 and 2." *Africa* 37: 50–71; 155–87.

Howes, Michael, and Robert Chambers. 1979. "Indigenous Technical Knowledge: Analysis, Implications and Issues." *IDS Bulletin* 10 (2): 5–11.

IWGIA (International Work Group for Indigenous Affairs). 2008. *The Indigenous World 2008*. Copenhagen, Denmark: IWGIA.

Kearney, Anne R. 1994. "Understanding Global Change: A Cognitive Perspective on Communicating through Stories." *Climatic Change* 27 (4): 419–41.

Knorr-Cetina, Karin. 1995. "Laboratory Studies: The Cultural Approach to the Study of Science." In *Handbook of Science and Technology Studies*, ed. Sheila Jasanoff, Gerald E. Markle, James C. Peterson, and Trevor J. Pinch, 140–66. Thousand Oaks, CA: Sage.

Kronik, Jakob. 2001. "Living Knowledge—Institutionalizing Learning Practices about Biodiversity among the Muinane and the Uitoto in the Colombian Amazon." PhD diss., Roskilde University, Roskilde, Denmark. http://rudar.ruc.dk/handle/1800/294.

Kronik, Jakob, and Dorte Verner, eds. Forthcoming. *Indigenous Peoples and Climate Change in Latin America and the Caribbean*. Washington, DC: World Bank.

Latour, Bruno. 1999. *Pandora's Hope—Essays on the Reality of Science Studies*. Cambridge, MA: Harvard University Press.

Latour, Bruno, and Steve Woolgar. 1986. *Laboratory Life: The Construction of Scientific Facts*. 2nd ed. Princeton, NJ: Princeton University Press.

Layton, Heather Marie, and Harry A. Patrinos. 2006. "Estimating the Number of Indigenous People in Latin America." In *Indigenous Peoples, Poverty, and Human Development in Latin America*, ed. Gillette Hall and Harry A. Patrinos, 25–39. New York: Palgrave.

Lévi-Strauss, Claude. 1962. *Den Vilde Tanke*. Danish translation of *Le pensée sauvage*, 1994. Copenhagen, Denmark: Gyldendal.

Mann, Charles C. 2005. *1491—New Revelations of the Americas before Columbus*. New York: Vintage Books.

McCarthy, E. Doyle. 1996. *Knowledge as Culture—The New Sociology of Knowledge.* London: Routledge.

Myers, Norman. 1984. "Eternal Values of the Parks Movement and the Monday Morning World." In *National Parks, Conservation and Development: The Role of Protected Areas in Sustaining Society,* ed. Jeffrey McNeely and Kenneth R. Miller, 656–60. Washington, DC: Smithsonian Institution Press.

Nepstad, Daniel, S. Schwartzman, B. Bamberger, M. Santilli, D. Ray, P. Schlesinger, P. Lefebvre, A. Alencar, E. Prinz, Greg Fiske, and Alicia Rolla. 2006. "Inhibition of Amazon Deforestation and Fire by Parks and Indigenous Lands." *Conservation Biology* 20 (1): 65–73.

O'Hara, Sarah L., Sarah E. Metcalfe, and F. Alayne Street-Perrott. 1994. "On the Arid Margin: The Relationship between Climate, Humans, and the Environment. A Review of Evidence from the Highlands of Central Mexico." *Chemosphere* 29 (5): 65–81.

O'Hara, Sarah L., F. Alayne Street-Perrott, and Timothy P. Burt. 1993. "Accelerated Soil Erosion around a Mexican Highland Lake Caused by Pre-Hispanic Agriculture." *Nature* 362: 48–51.

Orlove, Benjamin S., John C. H. Chiang, and Mark A. Cane. 2000. "Forecasting Andean Rainfall and Crop Yield from the Influence of El Niño on Pleiades Visibility." *Nature* 403: 68–71.

Parry, Martin L., Osvaldo F. Canziani, Jean Palutikof, Paul van der Linden, and Clair Hanson, eds. 2007. *Climate Change 2007: Impacts, Adaptation and Vulnerability. Contribution of Working Group II to the Fourth Assessment Report of the Intergovernmental Panel on Climate Change.* Cambridge, U.K.: Cambridge University Press.

Posey, Darrell Addison, and William Balée. 1989. *Resource Management in Amazonia: Indigenous and Folk Strategies.* Advances in Economic Botany, vol. 7. New York: New York Botanical Garden Press.

Rhoades, Robert F., and Virginia D. Nazarea. 1999. "Local Management of Biodiversity in Traditional Agroecosystems." In *Biodiversity in Agroecosystems,* ed. Wanda Williams Collins and Calvin O. Qualset, 215–36. Boca Raton, FL: CRC Press.

Robledo, Carmenza, Martin Fischler, and Alberto Patiño. 2004. "Increasing the Resilience of Hillside Communities in Bolivia: Has Vulnerability to Climate Change Been Reduced as a Result of Previous Sustainable Development Cooperation?" *Mountain Research and Development* 24 (1): 14–18.

Sahlins, Marshall. 1985. *Islands of History.* Chicago, IL: Chicago University Press.

Sale, Kirkpatrick. 1990. *The Conquest of Paradise: Christopher Columbus and the Colombian Legacy.* New York: Alfred A. Knopf.

Salinger, James, Mannava V. K. Sivakumar, and Raymond P. Motha. 2005. "Reducing Vulnerability of Agriculture and Forestry to Climate Variability and

Change: Workshop Summary and Recommendations." *Climatic Change* 70 (1/2): 341–62.

Shelter, Stanwyn. 1991. "Three Faces of Eden." In *Seeds of Change: A Quincentennial Commemoration,* ed. Herman J. Viola and Carolyn Margolis, 225–47. Washington, DC: Smithsonian Institution Press.

Sillitoe, Paul. 1998. "The Development of Indigenous Knowledge: A New Applied Anthropology." *Current Anthropology* 39 (2): 223–52.

Smit, Barry, and Johanna Wandel. 2006. "Adaptation, Adaptive Capacity and Vulnerability." *Global Environmental Change* 16: 282–92.

Snow, Dean R. 1995. "Microchronology and Demographic Evidence Relating to the Size of Pre-Columbian North American Indian Populations." *Science* 268: 1601–05.

Steward, Julian Haynes. 1937. *Basin Plateau Aboriginal Sociopolitical Groups.* Washington, DC: Smithsonian Institution Press.

Thomas, David S. G., Chasca Twyman, Henry Osbahr, and Bruce Hewitson. 2007. "Adaptation to Climate Change and Variability: Farmer Responses to Intra-Seasonal Precipitation Trends in South Africa." *Climatic Change* 83: 301–22.

Turner, Billie L. II, and Karl W. Butzer. 1992. "The Columbian Encounter and Land-Use Change." *Environment* 34 (8): 16–20, 37–44.

Usher, Peter J. 2000. "Traditional Ecological Knowledge in Environmental Assessment and Management." *Arctic* 53 (2): 183–93.

Verner, Dorte. Forthcoming. *Social Implications of Climate Change in Latin America and the Caribbean.* Washington, DC: World Bank.

Warren, D. Michael, Jan Slikkerveer, and David Brokensha, eds. 1991. *Indigenous Knowledge Systems: The Cultural Dimensions of Development.* London: Keegan Paul.

Warren, D. Michael, Jan Slikkerveer, and Sunday Titiola, eds. 1989. "Indigenous Knowledge Systems: Implications for Agriculture and International Development." Studies in Technology and Social Change 11. Technology and Social Change Program, Iowa State University, Ames, IA.

Watson-Verran, Helen, and David Turnbull. 1995. "Science and Other Indigenous Knowledge Systems." In *Handbook of Science and Technology Studies,* ed. Sheila Jasanoff, Gerald E. Markle, James C. Peterson, and Trevor J. Pinch, 115–39. Thousand Oaks, CA: Sage.

World Bank. 2000. *Can Africa Claim the 21st Century?* Washington, DC: World Bank.

———. 2002. *World Development Report 2002: Building Institutions for Markets.* Washington, DC: World Bank.

———. 2006. *World Development Report 2006: Equity and Development.* Washington, DC: World Bank.

PART II

SOCIAL JUSTICE AND CLIMATE ACTION

CHAPTER 7

Local Institutions and Adaptation to Climate Change

Arun Agrawal

The popular consensus on the reality of climate change, its human causes, and the severity of its impacts may not be very old, but most scholarly and policy literature holds that poor, natural resource–dependent, rural households will bear a disproportionate burden of adverse impacts (Mendelsohn et al. 2007; Kates 2000). Certainly, in many parts of the world, these effects already are in play, with potentially disastrous consequences for the poor (Adger et al. 2007; Adger, Arnell, and Tompkins 2005). But the rural poor[1] have successfully faced threats linked to climate variability in the past, even if climate change likely will increase the expected frequency and intensity of such threats (Mortimore and Adams 2001; Scoones 2001). Whether historically developed adaptation practices among the rural poor will be successful depends crucially on the nature of prevailing formal and informal rural institutions.[2]

Adaptation to climate change is highly local, and its effectiveness depends on local and extralocal institutions through which incentives for

I thank Neil Adger, Nilufar Ahmad, Simon Batterbury, Erin Carey, Mafalda Duarte, Gretel Gambarelli, Rachel Kornak, Julien Labonne, Maria Lemos, Catherine McSweeney, Robin Mearns, Daniel Miller, Andrew Norton, Jesse Ribot, and Paul Siegel; participants at the "Are There Limits to Adaptation?" conference held in February 2008 in London; and participants at the Social Dimensions of Climate Change workshop, organized by the World Bank Social Development Department and held in March 2008 in Washington, DC. Special thanks are due to Nicolas Perrin, who has been involved with the research and revisions for this chapter along every step.

individual and collective action are structured. Not only have existing institutions affected how rural residents responded to environmental challenges in the past, but they are also the fundamental mediating mechanisms that will translate the impact of external interventions to facilitate adaptation to climate change in the future. Institutional arrangements structure risks and sensitivity to climate hazards, facilitate or impede individual and collective responses, and shape the outcomes of such responses. Understanding how they function in relation to climate and its impacts, therefore, is a core component in designing interventions that can influence the adaptive capacity and adaptation practices of poor populations positively.[3]

Historical experience and knowledge about adaptation possibilities are crucial to future policy formulation regarding adaptation. This is because the specific nature of climate change impacts continues to be uncertain (especially for small territorial units),[4] even though it is evident that the general impacts of climate change will be striking and long lasting if current trends continue. Future efforts to address climate change and craft strategic initiatives to enhance the rural poor's adaptive capacity, therefore, can profitably examine historical adaptive responses, their institutional contexts and correlates, and the role of institutions in facilitating adaptation.

This chapter assesses the role of institutions by proposing an analytical classification of historically observed adaptation practices. It then uses the familiar distinctions among public, civic, and private domains to survey important recent work on adaptation; and outlines a framework through which to view the relationships among adaptation to climate change, livelihoods of the rural poor, and the role of institutions in facilitating external support for adaptation.[5] The institutions, adaptation, and livelihoods framework illustrated in figure 7.1 shows the critical role that institutions play in climate adaptation. Institutions structure the impacts of climate risks on households in a given ecological and social context; and they shape the degree to which households' responses are likely to be oriented, individually or collectively. They also mediate the influence of any external interventions on adaptation practices.

The argument related to institutions, adaptation, and livelihoods is based on comparative static analysis, where changes in institutions themselves as a result of climate change are not explicitly taken into account. Certainly, institutions are not static entities; they are likely to change even more as a result of the changing calculations of advantage and the nature of political interactions among relevant decision makers than because of climate impacts—but the chapter focuses more on their mediating role and

Figure 7.1. Institutions, Adaptation, and Livelihoods Framework

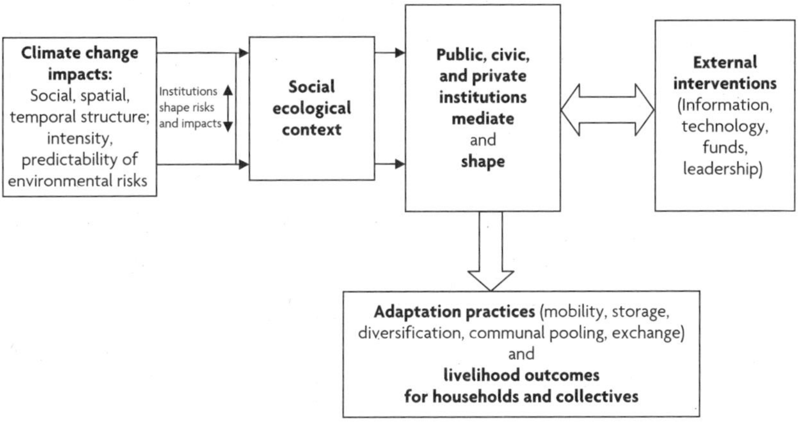

| Climate change impacts: Social, spatial, temporal structure; intensity, predictability of environmental risks | Institutions shape risks and impacts | Social ecological context | Public, civic, and private institutions mediate and shape | External interventions (Information, technology, funds, leadership) |

Adaptation practices (mobility, storage, diversification, communal pooling, exchange) and **livelihood outcomes for households and collectives**

Source: Author's illustration.

impacts on outcomes than on how climate changes will also affect local institutions. That question—how climate hazards of different kinds will affect institutions—is important, but it is not the subject of this chapter.

After briefly examining the relationships among climate-related vulnerabilities, adaptation practices, institutions, and external development interventions in the second section, the chapter develops a typology of adaptation practices (third section). It then applies the typology to the extensive data set on coping and adaptation strategies, generated by the United Nations Framework Convention on Climate Change (UNFCCC) (fourth section). The analysis of the 118 cases of adaptation drawn from the UNFCCC database permits three conclusions about climate adaptation and the role of institutions in adaptation, particularly for rural contexts: (1) local institutions play a central role in all observed adaptation efforts and practices, (2) civil and public sector organizations are key to local adaptations, and (3) private sector and market forces have been less important to adaptation in the studied cases. These findings, their explanation, and their implications are discussed in the concluding section of the chapter.

Climate Change, Vulnerability, and Adaptation

Consider a familiar example. A climate-related shock to livelihoods—for example, a drought in a semiarid region—has the potential to devastate the livestock owned by a household. Development interventions that increase the milk or meat yields from herds without increasing their capacity to survive in the face of fluctuations in fodder availability may increase total yield for the herd owner, but fail to smooth fluctuations across time periods. In the same agroecological context, privatization of land parcels can increase tenure security and encourage landowners to invest in the improvement of territorial infrastructure. But improvements may yield indifferent returns because of spatial and temporal fluctuations in rainfall that exacerbate household- and community-level vulnerability. However, if land is under open access in dry seasons, livestock-owning households can migrate to take opportunistic advantage of areas where forage is available—indeed, this is the strategy many of them use in drier areas of western India, Mongolia, and sub-Saharan Africa (Agrawal 1999). On the other hand, development of drought-resistant breeds of cattle and land tenure regimes that permit mobility may lead to lower overall output in terms of fodder, milk, or meat, but also may go together with greater capacity to withstand climatic variability.

Climate Vulnerability

Among the most important impacts of climate change on poor and vulnerable people are greater variability in temperatures and precipitation over time and across space and the impacts of such variability across asset types and households. With increasing climate variability, development interventions that do not attend to vulnerability, adaptive capacity, and resilience may worsen the circumstances of those they seek to benefit. Efforts to address vulnerability of the poor and to improve adaptive capacity require deeper attention to institutions at multiple scales and careful planning to ensure that institutions can work to help poorer groups who are most at risk from increasing volatility in climate phenomena and its human impact. These groups live close to subsistence margins, and variations in earnings and livelihoods capabilities are far more likely to plunge them below the margin, compared with those relatively well-off people who can draw on a variety of capital assets and institutional networks in times of stress.

In considering climate impacts, it is necessary to attend to their frequency, periodicity, intensity, and timing to understand how they affect adaptive capacity. The dynamics of risk exposure can be crucial in determining both the sensitivity and adaptive capacity of social groups and households. Repeated and unpredictable exposure to risks can drastically reduce the ability of even those households with high adaptive capacity to cope or respond effectively to risks. Thus, the adaptive capacity of a household or community may be depleted significantly as a result of a recent major shock to its livelihood and assets; similarly, households and communities that face regular occurrences of particular climate hazards are more likely to have developed adaptive responses over time, as long as the scale of the hazard is not great enough to wipe them out of existence.

In terms of intensity, frequency, regularity, and predictability, the dynamics of climate impacts are evidently related to the vulnerability of groups experiencing them. Irregularly and unpredictably repeated high-intensity environmental shocks will have the worst impacts on household- and community-level vulnerability. But vulnerability is also a function of the nature and types of assets that households and individuals possess. Human capital in the form of training, skills, and knowledge; social capital in terms of relationships and institutional access; financial capital in terms of liquid and nonliquid assets; natural capital in terms of available natural resources; and built capital in terms of infrastructure resources can all reduce the vulnerability of different social groups and households to climate variability and change-related impacts. In that sense, most adaptation choices that households and communities make depend on the nature and combination of assets and opportunities to which they have access.

Most recent studies on climate change have drawn on earlier work on vulnerability (Bohle et al. 1993; Watts and Bohle 1993), and highlight the fact that vulnerability to climate change is a function not only of biophysical outcomes related to variations and changes in temperature, precipitation, topography, and soils; but also of sociopolitical and institutional factors that can vary significantly at a relatively fine scale (Adger 2006). Structural and group characteristics such as gender, caste, race, ethnic affiliation, indigenity, and age—even when they are not consistent predictors—are often related closely with vulnerability. The degree to which they are associated with vulnerability tends to depend on location- and culture-specific factors; so, although climate change is a global phenomenon, adaptation to climate impacts inevitably and unavoidably is local (Ribot 1995; Blaikie et al. 1994).

Climate Vulnerability and Institutions

Although households and communities historically have used many different strategies to adapt to climate variability and the vulnerability resulting from it, their capacity to adapt depends in significant measure on the ways institutions regulate and structure their interactions, both among themselves and with external actors. All efforts to adapt depend for their success on specific institutional arrangements because adaptation never occurs in an institutional vacuum. Property rights and other institutions regulate access to resources and exposure to risks. Institutionalized monitoring and sanctioning in cases of individual or collective infractions of existing institutional rules are crucial for effective institutional functioning. Institutional rules about information sharing—even when informal—often help reduce vulnerability by facilitating ex ante planning an'd action. No wonder many indigenous pastoralist systems developed strong norms of information storage and exchange regarding spatial and temporal variability in precipitation and range productivity (Nyong, Adesina, and Elasha 2007; Agrawal 1999).

Indeed, the role of institutions at multiple scales, including local contexts, is accepted broadly in many analyses of climate and adaptation (Young and Lipton 2006; Batterbury and Forsyth 1999). But relatively little of the existing work has undertaken a careful or systematic analysis of the different types of institutions relevant to different forms of climate hazards–related adaptation, the different roles of local institutions in the context of adaptation, or the features of institutions that are most important for successful adaptation in rural contexts in the developing world (however, see Bakker 1999).

Types of Local Institutions. In examining the role of local institutions in facilitating adaptation, the familiar threefold classification of civic, public, and private institutions—primarily in their formal (and where relevant, informal) form—serves as a useful starting point. In many contexts, formal local institutions and organizations work in ways that promote informal processes, and these interactions can be vital to adaptation. Furthermore, although the analytical distinctions among these different types of organizations are important to bear in mind, in their functioning these organizations often enter into partner relationships and promote cross-domain collaborations.

There are strong reasons to believe that such partnerships among civic, public, and private organizations can prove extremely important in addressing climate hazards–related adaptation. Figure 7.2 proposes a schematic representation of such partnerships and collaborative arrangements as a first step in analyzing how institutions in the three spheres can work jointly to help facilitate adaptation.

Institutional partnerships have become especially common in the environmental arena and in the context of development projects. In many

Figure 7.2. Schema of Collaborative Institutional Arrangements for Environmental Action in the Context of Climate Change

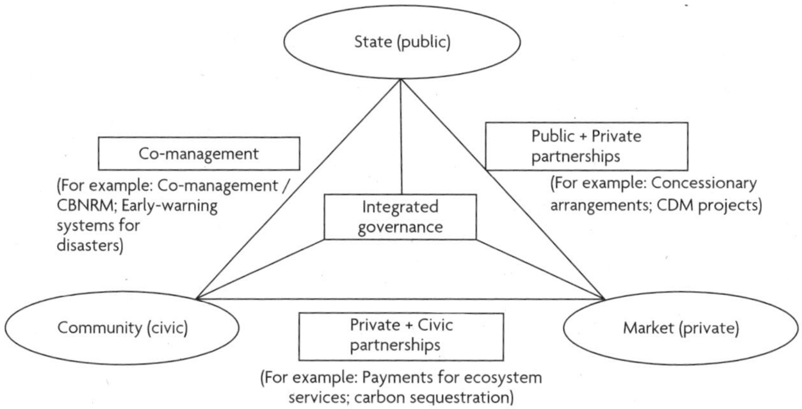

Source: Agrawal and Lemos 2007.
Note: CBNRM = Community-Based Natural Resource Management; CDM = Clean Development Mechanism.

instances, government agencies have sought to manage resources more effectively by partnering with civic bodies; to reduce pollution by working with corporations; to implement development projects in partnership with nongovernmental organizations; or to decentralize control over administrative functions and outsource important functions related to accounting, recordkeeping, financial management, and project monitoring and evaluation.

Figure 7.2 thus not only suggests the possibility of institutional partnerships across the civic-public-private domains in the context of climate adaptation, but also highlights the importance of such partnerships. A number of people have called climate change one of the greatest market failures of human history. It is clear that adaptation to climate change will require the concerted efforts of decision makers in diverse institutions across multiple scales.

Roles of Local Institutions. Broadly speaking, different local institutions shape the effects of climate hazards on adaptation and livelihoods in three important ways:

1. Local institutions structure environmental risks and variability, and thereby the nature of climate impacts and vulnerability.
2. They create the incentive framework within which outcomes of individual and collective action unfold.

3. They are the media through which external interventions reinforce or undermine existing adaptation practices.

Local institutions play a crucial role in influencing the ex ante adaptive capacity of communities and the adaptation choices made ex post by community members. The above example also shows the importance of close connections between local and higher-level institutions, and the extent to which such connections enable rural residents to leverage their membership in local institutions for gains from outside the locality. Indeed, the critical role of institutions is underscored repeatedly in studies of adaptive capacity and adaptation choices (Ivey et al. 2004; Adger 1999).

Local institutions structure livelihood impacts of climate hazards through a range of indispensable functions they perform in rural contexts. Institutional functions include information gathering and dissemination, resource mobilization and allocation, skills development and capacity building, leadership, and means of connecting with other decision makers and institutions. Each of these functions can be disaggregated further, but the extent to which any given institution performs the above functions depends largely on the objectives with which the institution was formed, and on the problems it has come to address over the course of its existence.

Climate Impacts and Types of Adaptation

Two relatively recent major surveys of climate change and its impacts have identified many areas in which there now is significant scientific consensus substantiating claims of the adverse impacts of climate change on agricultural, food, water, social, and ecological systems (Parry et al. 2007; Solomon et al. 2007; Stern 2006). There also is a well-developed body of work around the key concepts of vulnerability, resilience, and adaptation in the context of global environmental change. This evidence suggests climate change will stress existing livelihood options and, even more important, will make them more unpredictable because of the increased volatility in climate impacts (Yohe and Tol 2002; Rosenzweig and Parry 1994).

The problem of increased volatility and risk resulting from climate change is especially important. It means that many more vulnerable households periodically can be driven into destitution and hunger, and can find it difficult to recover. This is because, by definition, the incomes and livelihoods of poorer, more vulnerable households are closer to the lines between adequate subsistence, malnutrition, and starvation. The role of rural local institutions in this regard is crucial. Not only do institutions influence how households are affected by climate impacts; they also shape the ability of households to respond to climate impacts and to pursue different adaptation practices, and they mediate the flow of external interventions in the context of adaptation. The nature of access of different households and social groups to institutions and institutionally allocated resources is a critical factor in their ability to adapt successfully.

It is clear, therefore, that development strategies and institutional interventions that focus simply or even mostly on improving aggregate benefits to poor households, without taking into account how households can address fluctuations in their livelihoods, are ill suited to address the impacts of climate change. There are two reasons why they are ill suited. Such initiatives ignore a key feature of climate-related stresses: they will make livelihoods more precarious. And interventions and strategies that focus on increasing aggregate benefits to poor households ignore rural poor people's very real concerns about preventing starvation and destitution. Indeed, a long tradition of scholarship in the social sciences has argued about the extent to which many rural households live close to the margins of subsistence (Scott 1976; Wolf 1969) and seek to avoid drops in livelihoods.

To strengthen the adaptive capacity of the rural poor population, governments and other external actors also need to understand, take advantage of, and strengthen already existing strategies that many households and social groups use singly or collectively. In different parts of the world, many rural communities already experience high levels of climate variability, and they have developed more or less effective responses to address such variability. Much of the Sahelian region, for example, faces extreme irregularity in rainfall, with recurrent droughts. A number of scholars have argued, based on available data, that annual rainfall levels in the region have declined together with increases in interannual and spatial variability and the intensity of drought events (Tarhule and Lamb 2003; Hulme et al. 2001). In response, farmers have adapted their farming, livestock-rearing, and other income-generating activities to achieve some degree of sustainability in their livelihoods (Nyong, Adesina, and Elasha 2007; Blanco 2006).

Increases in environmental risks as a result of climate change can be classified in many ways: short-term versus long-term, those resulting from sudden disasters versus those resulting from slow but secular changes in trends, predictable versus unpredictable, and the like. In looking at household livelihood strategies, a particularly useful way to think about climate-related risks is to examine how they affect livelihood capabilities over time, across space, across asset classes, and across households. Those four types of risks to livelihood comprise the major conceptual categories of the ways variability threatens the ability of households to secure livelihoods. Thinking of adaptation in relation to these four forms of climate risks thus enables an analytically connected approach to classifying adaptation practices.

The basic adaptation strategies in the context of climate and other environmental risks to livelihoods can be linked to the following five analytical risk management categories:

1. mobility—the distribution of risk across space
2. storage—the distribution of risk across time
3. diversification—the distribution of risk across asset classes
4. communal pooling—the distribution of risk across households
5. market exchange—the purchase and sale of risk via contracts, which may substitute for any of the other four categories when households have access to markets (Halstead and O'Shea 1989).

The effectiveness of these five classes of adaptation practices is partly a function of the social and institutional contexts in which they are pursued. Where successful, these responses pool uncorrelated risks associated with flows of benefits from different classes of assets owned by households and economic agents. They also can enable a shift away from more risky economic strategies to other, less risky ones.

The above classification of adaptation practices is different from other classifications that view adaptation as proactive or reactive, individual or collective, spontaneous or planned. Although such other ways of thinking about adaptation are useful, they are not related to the basic types of risks that climate hazards pose; and, therefore, they are analytically fuzzier than the classification proposed here. The classification system proposed here also undermines the often-proposed distinction between coping and adaptation,[6] which essentially is dependent on the extent to which a given response to a climate hazard produces long-lasting effects on adaptive capacity. When climate hazards are repeated, the distinction between short-term and long-term adaptation (that is, coping versus supposedly real adaptation) breaks down, as does that between proactive and reactive adaptation. Furthermore, the fivefold classification of adaptation practices is equally relevant to coping and adaptation because both are intended to address environmental risks and stresses.

Mobility, Storage, Diversification, Pooling, and Exchange

Mobility is perhaps the most common and seemingly natural response to environmental risks. It pools risks across space, and is especially successful in combination with clear information about the spatial and temporal distribution of precipitation. It is especially important as an adaptation practice for agropastoralists in sub-Saharan Africa, west and south Asia, and most dry regions of the world (Niamir 1995).

In the context of climate change, mobility sometimes has been viewed as a maladaptation, in which climatic stresses lead to involuntary migrations on a massive scale, with attendant social and political instabilities. This view is especially prevalent in many policy briefings and papers (Purvis and Busby 2004; Schwartz and Randall 2003). However, mobility also is a way of life for large groups of people in semiarid regions, and a long-standing mechanism to deal with spatiotemporal variations in rainfall and range productivity. Therefore, whether mobility is considered a desirable adaptation often depends on the status of the social groups in question.

For agricultural populations, mobility often can be the last resort in the face of environmental risks and disruption of livelihoods (McGregor 1994). For pastoralist and agropastoralist populations, however, efforts to limit mobility could lead to greater vulnerability and lower adaptive capacity (Davies and Bennett 2007; Agrawal 1999). At the same time, frequent movement of people with their animals raises particularly intricate questions about the role of institutions in facilitating adaptation. Most governance institutions are designed with sedentary populations as their target groups. To address the needs of mobile populations, the role of information in tracking human and livestock movements, and the mobile provision of basic services such as health, education, credit, and marketing of animal products, are especially important to reinforce adaptive capacity.

Storage pools and reduces risks across time. When combined with well-constructed infrastructure, low levels of perishability, and a high level of coordination across households and social groups, it is an effective measure against even complete livelihood failures. As an adaptation practice to address risks, storage is relevant to individual farmers and communities to address food as well as water scarcities. Indeed, in light of the significant losses of food and other perishable commodities all over the developing world, improvements in storage technologies and institutions have immense potential to improve rural livelihoods.

Diversification pools risks across the assets and resources of households and collectives. Highly varied in form, it can occur in relation to productive and nonproductive assets, consumption strategies, and employment opportunities. It is reliable to the extent that benefit flows from assets are subject to uncorrelated risks (Ellis 2000; Behnke, Scoones, and Kerven 1993; Sandford 1983). Diversifying households typically give up some returns in exchange for the greater security provided by diversification. Davies and Bennett (2007) provide a striking example from the Afar pastoralists of Ethiopia. Many of them would be willing to live with some level of poverty in exchange for reduction in vulnerability.

Communal pooling is an adaptation practice that involves joint ownership of assets and resources; sharing of wealth, labor, or incomes from particular activities across households; and/or mobilization and use of resources that are held collectively during times of scarcity. It pools risks across households. This practice is most effective when the benefits from assets owned by different households and the livelihood benefit streams are uncorrelated. When a group is affected in a similar manner by adverse climate hazards—for example, floods or drought—communal pooling is less likely to be an effective response.

Although communal pooling can occur in combination with mobility, storage, and diversification, its hallmark is joint action by members of a group with the objective of pooling their risks and resources. Joint action increases the range of impacts with which collected households might have to deal, compared with the range of impacts individual households could have encountered. It also requires functioning and viable institutions to coordinate activities across households. It is one way for social groups—especially those dependent on natural resources for their livelihoods—to enhance their capacity to adapt to the impacts of future climate change (Adger 1999).

Market exchange may be the most versatile adaptation response. Indeed, markets and exchanges are a characteristic of almost all human groups; and they are a mechanism not only for adaptation to environmental risks, but also for specialization, trade, and welfare gains that result from specialization and trade at multiple scales. Market exchange–based adaptation practices can substitute for the other four practices when rural poor people have access to markets. But they are likely to do so mainly when there are well-developed institutions to facilitate market access. Furthermore, equity in adaptation practices based on market exchanges typically requires significant attention to the institutional means through which access to markets and market products becomes available to households. In the absence of institutional mechanisms that can ensure equity, the rural poor are less likely to benefit from purely market exchange–based adaptation.

Influences on Adaptation Practices

The choice of specific adaptation practices is dependent on social and economic endowments of households and communities, as well as their ecological location, networks of social and institutional relationships, institutional articulation and access, and the availability of resources and power. For example, the poor are more likely to migrate in response to crop failure; the rich are more likely to rely on storage and exchange. This is because the rich are more likely to have institutionally secure access to resources that make forced migration unnecessary. Migration is more likely to be an effective long-term strategy for pastoralists and agropastoralists confronting lower rainfall or range productivity, in contrast to settled agriculturists. However, the ability to migrate depends on the nature of property institutions over pasturelands along the migration route.

Similarly, whether households and communities can diversify into new occupations and assets depends not only on their ability to trade some level of returns for lowered risks, but also on access to capital, availability of skills training, and the effectiveness of agricultural extension institutions. The importance of institutions as the scaffolding on which households and individuals can coordinate their expectations and thereby create effective collective action has been demonstrated repeatedly. The success of market exchanges also depends on institutions able to reduce or eliminate problems of adverse selection, moral hazard, and burdensome transaction costs.

In addition, the different adaptation practices discussed above have natural affinities and incompatibilities. Storage and mobility tend not to go together. Other combinations complement each other: storage and exchange can play temporal variability against spatial variability (Halstead and O'Shea 1989). Diversification similarly enables agricultural households simultaneously to reduce risks and reap the benefits of market exchange.

Adaptation, Institutions, and Livelihoods Framework: Case Evidence

This section uses evidence from the UNFCCC Coping Strategies Database to assess comparatively the role of local rural institutions in facilitating adaptation. The database provides a useful review and summary of 118 cases of adaptation worldwide.[7] These cases form an empirical basis for examining the distribution of adaptation practices, the role of local institutions in facilitating adaptation, and the means by which institutions mediate between external interventions and improvements in local adaptive capacity.

The database on coping strategies contains many different types of cases, distributed across 46 countries, with the preponderance of the cases from Africa (45) and Asia (58). Although it is called the Coping Strategies Database, it primarily includes examples that are what might conventionally be thought of as adaptations to environmental variability—the use of particular labor-sharing techniques in agriculture, pooling of assets across households, use of different types of foods during different times of the year, building of local infrastructure, and the like. The inclusion of adaptation strategies in a database titled "Coping Strategies" shows the difficulty of distinguishing empirically between the long-term and short-term responses.

The information on different adaptation and coping practices is contained as a description of each case. Thus, the UNFCCC's database is not a searchable spreadsheet of information on variables for each case; rather, it is a set of documents and descriptions for each case. The cases were collected by UNFCCC's researchers, both as contributions from in-country experts and from the existing literature. The database was not designed to collect only cases illustrating the role of institutions in adaptation or, indeed, to favor collection of cases that exemplified the role of a particular type of institution—civic, private, or public. Instead, the database and its cases combine in a single location the wealth of historical experience on adaptation and coping strategies.

For the ensuing analysis in the chapter, all cases in the database were analyzed to derive the relevant information on adaptation strategies, their relationship to different kinds of local and external institutions, climate hazards, and types of external assistance available in a given case for adaptation and coping. The analysis of the strategies identified and discussed in the 118 cases shows that they can be classified either as individual illustrations or examples of combinations of the five different classes of adaptation practices described earlier: mobility, storage, diversification, communal pooling, and market exchange.

The evidence in the cases presents some useful, even provocative patterns. Perhaps the most interesting ones concern the near-complete absence of mobility in the examined cases (see table 7.1), and the occurrence of exchange only in combination with at least one other type of adaptation

Table 7.1. Frequency Distribution of Different Classes of Adaptation Practices (n = 118)

Class of Adaptation Practice	Corresponding Adaptation Strategies	Frequency
Mobility	1. Agropastoral migration 2. Wage labor migration 3. Involuntary migration	2
Storage	1. Water storage 2. Food storage (crops, seeds, forest products) 3. Animal/livestock storage 4. Pest control	11
Diversification	1. Asset portfolio diversification 2. Skills and occupational training 3. Occupational diversification 4. Crop choices 5. Production technologies 6. Consumption choices 7. Animal breeding	33
Communal pooling	1. Forestry 2. Infrastructure development 3. Information gathering 4. Disaster preparation	29
Market exchange	1. Improved market access 2. Insurance provision 3. New product sales 4. Seeds, animals, and other input purchases	1
Storage + diversification	Examples of combinations of adaptation classes are drawn from the strategies listed above.	6
Storage + communal pooling		4
Storage + market exchange		5
Diversification + communal pooling		4
Diversification + market exchange		25
Unidentified		2
Total		122

Source: UNFCCC Coping Strategies Database.
Note: UNFCCC = United Nations Framework Convention on Climate Change. Frequency totals 122 (rather than 118) because four cases were instances of more than two forms of adaptation, occurring within the same case.

practice. It makes intuitive sense that, as an adaptation practice, market exchange should be possible only when households and communities also have adopted other adaptation practices to make something available for exchange. Table 7.1 also suggests that the most common classes of adaptation responses are diversification and communal pooling on their own, and diversification plus exchange as a pair.

The data also show two other interesting patterns:

1. In nearly all cases, local institutions are required to enable households and social groups to deploy specific adaptation practices

(see table 7.2). In 77 cases, the primary structuring influence for adaptation flows from local institutions. In 41 cases, local institutions work in conjunction with external interventions. Given that the cases included in the database were collected without regard to the role of institutions, the structuring importance of local institutions in the coping and adaptation cases shows, at a minimum, that they play a key role in adaptation at the local level. Even if one started with a prior assumption that institutions have no role to play in adaptation, and considered the positive evidence about the important role of local institutions in nearly all the available cases, one would be forced to revise the initial belief substantially.

2. In all cases where external support is present, it is channeled through local formal and informal institutions to enable adaptation. The inference is evident: without local institutions, rural poor groups will find it far costlier to pursue any adaptation practice relevant to their needs.

Given the importance of institutions to adaptation practices, it is critical to attend to three issues. The first issue concerns the distribution of institutional types (civic, public, and private) in facilitating local adaptation. The second issue concerns how different types of institutions relate to different classes of adaptation practices. The third issue concerns the importance of understanding the distribution of different types of institutions in relation to their mediating role for external interventions. Figure 7.3, table 7.2, and

Table 7.2. Combinations of Adaptation Practices and Institutions in the UNFCCC Database (n = 118)

Practice	Institutions					
	Civic	Public	Private	Public + Civic	Private + Civic	Total
Storage	8	0	0	3	0	11
Diversification	**19**	0	1	**12**	1	33
Communal pooling	**11**	4	0	**14**	0	29
Storage + diversification	2	0	0	2	0	4
Storage + exchange	4	0	0	1	1	6
Diversification + exchange	**13**	0	**4**	5	**4**	26
Other	4	2	0	3	0	9
Total	61	6	5	40	6	118

Source: UNFCCC Coping Strategies Database.
Note: UNFCCC = United Nations Framework Convention on Climate Change. Boldface type shows important patterns in the data.

Figure 7.3. Formal versus Informal Institutions in Adaptation

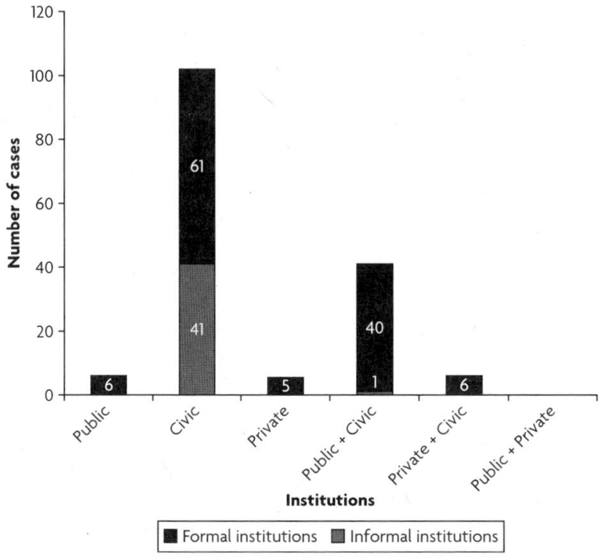

Source: UNFCCC Coping Strategies Database.
Note: UNFCCC = United Nations Framework Convention on Climate Change.

table 7.3 provide an initial assessment in regard to these three questions, on the basis of the UNFCCC data.

Figure 7.3 provides a summary of the nature of institutional involve-ment in adaptation practices at the local level. The data make evident two points. The first point is that civic and public-plus-civic institutions are the ones most commonly involved in facilitating adaptation to climate change (in more than 80 percent of the included cases). Private or market insti-tutions have played a role in facilitating or reinforcing adaptation in a very small proportion (less than 10 percent) of the cases. This finding is simultaneously a challenge and an opportunity to identify ways of creating incentives and partnerships involving the private sector and market actors more intimately in facilitating adaptation.

Another salient pattern in the data is that local-level civic institutions, when functioning on their own, often tend to be informal institutions. How-ever, when public institutions are involved in adaptation practices, their relationship far more often is with formal civic institutions. (See the distri-bution of formal and informal institutional arrangements for adaptation

Table 7.3. Local Institutions and Their Mediation of External Interventions to Promote Adaptation (n = 41)

Intervention	Institutions				
	Civic	Public	Public + Civic	Private + Civic	Total
Information	2	0	8	0	10
Technical input	4	2	1	0	7
Financial support	0	2	6	1	9
Information + technical inputs	4	0	2	0	6
Technical input + financial support	4	0	1		5
Other	2	0	2	0	4
Total	16	4	20	1	41

Source: UNFCCC Coping Strategies Database.
Note: UNFCCC = United Nations Framework Convention on Climate Change.

as reflected in the bars for civic and public-plus-civic institutions in figure 7.3). One of the implications of these data is that there are potentially significant gains to be made by identifying ways of encouraging informal processes through formal interventions to facilitate adaptation and greater adaptive capacity.

The second point made evident by the data concerns how different types of institutions correlate with particular combinations of adaptation practices. The UNFCCC data does not provide detailed evidence on the subject, but it is possible to generalize in a preliminary way, based on its information about how civic, public, and private rural institutions connect with different classes of adaptation practices (see table 7.2).

Although the UNFCCC database does not provide enough information to make a detailed assessment of the subdivisions within the broad categories of civic, public, and private institutions, it does suggest that public and market institutions do not promote mobility;[8] that public institutions only infrequently are associated with market exchange processes promoting adaptation; and that when market actors are involved in adaptation practices, it is likely that they will assist exchange-based efforts.

It also is clear that much of the institutional action is focused around civic and a combination of public and civic institutions. A few points are still worth highlighting from this information. The first point is that civic institutions and partnerships between civic and public institutions seem to occur more frequently to promote diversification and communal pooling. There are relatively few instances of civic institutions promoting storage, mobility, or a combination of different adaptive practices. In contrast, much of the involvement of private institutions and the partnership between civic and private institutions seems to focus on the promotion of diversification and market exchange.

Table 7.3 provides a summary overview of how civic, public, and private institutions mediate external interventions to promote adaptation. It focuses on the 41 of 118 cases in the data set that clearly show the involvement of external actors in promoting adaptation. The information in table 7.3 suggests that the major external interventions to support local adaptation efforts have focused on providing information and financial support. There are fewer cases in which a variety of external interventions have been combined to facilitate adaptation, and in no case have external actors provided strong leadership or attempted local institutional reconfiguration to support adaptation. A closer look at the data explains these patterns. The vast majority of cases of information provision and financial support concern adaptation practices related to disaster preparedness, early-warning systems about rainfall failure, and private or public infrastructure that could withstand climate hazards such as floods and storms. The conclusion is inescapable: external forms of support focus on an incredibly small slice of the huge range of adaptation mechanisms that local actors and institutions are inventing and attempting.

The UNFCCC data allow several inferences concerning the distribution of five adaptation practices (as discussed earlier in this chapter), the relationship between adaptation practices and local institutions, and the relationship between different types of local institutions and how they mediate external interventions to facilitate adaptation. Three of the more important implications of these data are worth reiterating:

1. Local institutions are crucial to the successful pursuit of local and externally facilitated local adaptation practices. Examples of coping and adaptation were included in the UNFCCC database without regard to the role institutions played in adaptation. Nonetheless, institutions turn out to play an important structuring role in the cases in the database. It is likely, therefore, that local institutions, in general, play a key role in adaptations by households.

2. Civic and informal institutions are key mechanisms to achieve most forms of adaptation. They play an extremely important role in adaptation in a variety of ecological and social contexts; and they do so particularly for diversification, communal pooling, and the combination of diversification and exchange.

3. Although the available data do not possess sufficient detail to make fine distinctions about the characteristics of institutions that are most

important in pursuing adaptation, they still suggest the highly under-exploited strengths of the private sector and market forces in helping enhance adaptive capacity in marginal environments. Private sector institutions are present in only 10 percent of the cases included in the database, despite the fact that exchange is a form of adaptation involved in nearly 30 percent of the cases there. The strong role private sector organizations can play in adaptation, therefore, may need to be realized through policy interventions that encourage them to take such a role.

Conclusion

Impacts of ongoing climate change will greatly increase the vulnerability of poorer, more marginalized households in developing countries. The planetary scale of climate change notwithstanding, its impacts will be highly differentiated spatially—increasing average temperatures will hide a diversity of impact variations on regions, communities, and households; and similarly across annual, seasonal, and diurnal ranges. This will occur because vulnerability to climate change is determined socially and institutionally, even when triggered biophysically. For that reason, adaptation to the inevitable effects of climate change is unavoidably local, and local rural institutions have a critically important role in promoting effective adaptation and in enhancing the adaptive capacity of vulnerable rural populations.

Despite the critical importance of rural institutions in shaping the adaptive responses of humanity to climate change and variability, the literature on the subject is in its infancy. This chapter has reviewed the state of knowledge on the role and importance of institutions in adaptation to propose a new classification for adaptation practices, based on the underlying forms of risks that will be exacerbated by climate change; and it has used data on coping and adaptation practices in the UNFCCC Coping Strategies Database to generate three key findings.

The first finding is that, although unavoidably local, adaptation always occurs in an institutional context. Rural institutions are crucial in shaping adaptation and its outcomes. They are key mediating bodies that connect households to local resources, determine how flows of external support will be distributed among different social groups, and link local populations to national policies and interventions. The second finding is that, in ongoing forms of adaptation, civic and public-plus-civic institutional partnerships are the most common ones. These arrangements are associated in particular with the promotion of diversification and communal pooling and, in some instances, with the promotion of combined diversification and exchange-related adaptations. Finally, private sector organizations and institutions have played only a limited role in local adaptation efforts; even in adaptations where market exchange is a key element, private sector institutions are present only infrequently. New incentives and policy interventions will be necessary if private organizations and institutions are to become involved more centrally in facilitating adaptation.

The findings of this chapter have a number of implications for incorporating local institutions more closely in the context of adaptation to climate change. First, more effective adaptation likely requires greater support for institutional partnerships. Such partnerships are crucial to local adaptation. More specifically, partnerships between local public and civil society institutions are associated more closely with adaptation practices related to diversification and communal pooling. Partnerships between private and civil society institutions are relatively uncommon and need encouragement. They tend to be more closely associated with market exchange and storage-based adaptation practices.

A second implication of the chapter's research is that local institutional capacities need to be enhanced for more effective adaptation. Because the nature and intensity of future climate impacts are likely to be more adverse than what existing climate variability indicates, external support is necessary to promote institutional capacities to support adaptation. Such support to local institutions, in itself, will constitute a form of adaptation to climate risks, even if not by households. Such improvements in the ability of institutions to support adaptation by households nonetheless are key elements for effective adaptation in the future.

Finally, it is fair to suggest that, given the relative newness of research on adaptation and institutions, improvements in existing knowledge and the development of greater adaptive capacity will require far greater investments in research and knowledge creation than previously have been available.

Notes

1. This chapter uses the term "rural poor" to refer to the more marginalized and disadvantaged, and often the most vulnerable, social groups in rural areas.

2. "Adaptation" here refers to the actions and adjustments undertaken to maintain the capacity to deal with stresses induced by current and future external changes (Nelson, Adger, and Brown 2007; Alland 1975). This broad definition covers the kinds of adjustments that various agents make in response to climate hazards. It is broad enough to cover adaptations that promote future adaptive capacity, leave it unchanged, or reduce it because the outcomes of specific adaptations cannot be predicted perfectly. Adaptation is a process variable.

3. "Adaptive capacity" refers to the preconditions that enable actions and adjustments in response to current and future external changes. These preconditions comprise both social and biophysical elements (Nelson, Adger, and Brown 2007).

4. Such uncertainty is primarily the result of the scale at which projections about climate change can be made through general circulation models, the main source of information about future changes in climate.

5. In focusing on both adjustment and coping strategies, the chapter broadly follows the definition of adaptation as used by the Intergovernmental Panel on Climate Change (Adger et al. 2007, pp. 719–20).

6. "Coping" refers to the short-term use of existing resources to achieve various desired goals during and immediately after adverse conditions of a hazardous event or process. The strengthening of coping capacities, together with preventive measures, is an important aspect of adaptation; and it usually builds resilience to withstand the effects of natural and other hazards.

7. The total number of discrete cases in the UNFCCC database is 138. However, a number of the cases essentially are duplications of information, especially in the water harvesting and forest sectors.

8. Note that the UNFCCC database does not provide cases of wage labor diversification and mobility or agropastoralist migration as instances of adaptation practices.

References

Adger, W. Neil. 1999. "Social Vulnerability to Climate Change and Extremes in Coastal Vietnam." *World Development* 27 (2): 249–69.

———. 2006. "Vulnerability." *Global Environmental Change* 16 (3): 268–81.

Adger, W. Neil, Shardul Agrawala, M. Monirul Qader Mirza, Cecilia Conde, Karen O'Brien, Juan Pulhin, Roger Pulwarty, Barry Smit, and Kiyoshi Takahashi. 2007. "Assessment of Adaptation Practices, Options, Constraints and Capacity." In *Climate Change 2007: Impacts, Adaptation and Vulnerability. Contribution of Working Group II to the Fourth Assessment Report of the Intergovernmental Panel on Climate Change*, ed. Martin L. Parry, Osvaldo

F. Canziani, Jean Palutikof, Paul van der Linden, and Clair Hanson, 717–43. Cambridge U.K.: Cambridge University Press.

Adger, W. Neil, Nigel W. Arnell, and Emma L. Tompkins. 2005. "Successful Adaptation to Climate Change Across Scales." *Global Environmental Change* 15: 77–86.

Agrawal, Arun. 1999. *Greener Pastures: Politics, Markets, and Community among a Migrant Pastoral People*. Durham, NC: Duke University Press.

Agrawal, Arun, and Maria Carmen Lemos. 2007. "A Greener Revolution in the Making? Environmental Governance in the 21st Century." *Environment* 49 (5): 36–45.

Alland, Alexander Jr. 1975. "Adaptation." *Annual Review of Anthropology* 4: 59–73.

Bakker, Karen, ed. 1999. "A Framework for Institutional Analysis." Working Paper 3, Societal and Institutional Responses to Climate Change and Climate Hazards, Environmental Change Institute, University of Oxford, Oxford, U.K.

Batterbury, Simon, and Tim Forsyth. 1999. "Fighting Back: Human Adaptations in Marginal Environments." *Environment* 41 (6): 6–11, 25–30.

Behnke, Roy H., Ian Scoones, and Carol Kerven, eds. 1993. *Range Ecology at Disequilibrium: New Models of Natural Variability and Pastoral Adaptation in African Savannas*. London: Overseas Development Institute.

Blaikie, Piers, Terry Cannon, Ian Davis, and Ben Wisner. 1994. *At Risk: Natural Hazards, People's Vulnerability and Disasters*. London: Routledge.

Blanco, Ana V. Rojas. 2006. "Local Initiatives and Adaptation to Climate Change." *Disasters* 30 (1): 140–47.

Bohle, Hans-Georg, Thomas E. Downing, and Michael J. Watts. 1993. "Climate Change and Social Vulnerability." *Global Environmental Change* 4 (1): 37–48.

Davies, Jonathan, and Richard Bennett. 2007. "Livelihood Adaptation to Risk: Constraints and Opportunities for Pastoral Development in Ethiopia's Afar Region." *Journal of Development Studies* 43 (3): 490–511.

Ellis, Frank. 2000. *Rural Livelihoods and Diversity in Developing Countries*. Oxford, U.K.: Oxford University Press.

Halstead, Paul, and John O'Shea, eds. 1989. *Bad Year Economics: Cultural Responses to Risk and Uncertainty*. Cambridge, U.K.: Cambridge University Press.

Hulme, Mike, Ruth Doherty, Todd Ngara, Mark New, and David Lister. 2001. "African Climate Change: 1900–2100." *Climate Research* 17 (2): 145–68.

Ivey, Janet L., John Smithers, Rob de Loe, and Reid D. Kreutzwiser. 2004. "Community Capacity for Adaptation to Climate-Induced Water Shortages: Linking Institutional Complexity and Local Actors." *Environmental Management* 33 (1): 36–47.

Kates, Robert W. 2000. "Cautionary Tales: Adaptation and the Global Poor." *Climatic Change* 45 (1): 5–17.

McGregor, JoAnn. 1994. "Climate Change and Involuntary Migration: Implications for Food Security." *Food Policy* 19 (2): 120–32.

Mendelsohn, Robert, Alan Basist, Predeep Kurukulasuriya, and Ariel Dinar. 2007. "Climate and Rural Income." *Climatic Change* 81 (1): 101–18.

Mortimore, Michael J., and William M. Adams. 2001. "Farmer Adaptation, Change, and 'Crisis' in the Sahel." *Global Environmental Change* 11 (1): 49–57.

Nelson, Donald, W. Neil Adger, and Katrina Brown. 2007. "Adaptation to Environmental Change: Contributions of a Resilience Framework." *Annual Review of Environment and Resources* 32: 395–419.

Niamir, Maryam. 1995. "Indigenous Systems of Natural Resource Management among Pastoralists of Arid and Semi-Arid Africa." In *The Cultural Dimension of Development: Indigenous Knowledge Systems,* ed. D. Michael Warren, L. Jan Slikkerveer, and David Brokensha, 245–57. London: Intermediate Technology Publications.

Nyong, Anthony, F. Adesina, and B. Osman Elasha. 2007. "The Value of Indigenous Knowledge in Climate Change Mitigation and Adaptation Strategies in the African Sahel." *Mitigation and Adaptation Strategies for Global Change* 12 (5): 787–97.

Parry, Martin L., Osvaldo Canziani, Jean Palutikof, Paul van der Linden, and Clair Hanson, eds. 2007. *Climate Change 2007: Impacts, Adaptation and Vulnerability. Contribution of Working Group II to the Fourth Assessment Report of the Intergovernmental Panel on Climate Change.* Cambridge, U.K.: Cambridge University Press.

Purvis, Nigel, and Joshua Busby. 2004. *The Security Implications of Climate Change for the UN System.* Washington, DC: Woodrow Wilson International Center for Scholars.

Ribot, Jesse. 1995. "The Causal Structure of Vulnerability: Its Application to Climate Impact Analysis." *GeoJournal* 35 (2): 119–22.

Rosenzweig, Cynthia, and Martin L. Parry. 1994. "Potential Impact of Climate Change on World Food Supply." *Nature* 367: 133–38.

Sandford, Stephen. 1983. *Management of Pastoral Development in the Third World.* Chichester, U.K.: John Wiley.

Schwartz, Peter, and Doug Randall. 2003. "An Abrupt Climate Change Scenario and Its Implications for United States National Security: Imagining the Unthinkable." U.S. Department of Defense, Washington, DC.

Scoones, Ian, ed. 2001. *Dynamics and Diversity: Soil Fertility and Farming Livelihoods in Africa.* London: Earthscan.

Scott, James C. 1976. *The Moral Economy of the Peasant: Rebellion and Subsistence in Southeast Asia.* New Haven, CT: Yale University Press.

Solomon, Susan, Dahe Qin, Martin Manning, Melinda Marquis, Kristen Averyt, Melinda M. B. Tignor, Henry LeRoy Miller Jr., and Zhenlin Chen, eds. 2007.

Climate Change 2007: The Physical Science Basis. Contribution of Working Group I to the Fourth Assessment Report of the Intergovernmental Panel on Climate Change. Cambridge, U.K.: Cambridge University Press.

Stern, Nicholas. 2006. *The Economics of Climate Change: The Stern Review.* Cambridge, U.K.: Cambridge University Press.

Tarhule, Aondover, and Peter J. Lamb. 2003. "Climate Research and Seasonal Forecasting for West Africans: Perceptions, Dissemination, and Use." *Bulletin of the American Meteorological Society* 84: 1741–59.

Watts, Michael, and Hans-Georg Bohle. 1993. "The Space of Vulnerability: The Causal Structure of Hunger and Famine." *Progress in Human Geography* 17 (1): 43–67.

Wolf, Eric R. 1969. *Peasant Wars of the Twentieth Century.* New York: Harper and Row.

Yohe, Gary, and Richard S. J. Tol. 2002. "Indicators for Social and Economic Coping Capacity: Moving toward a Working Definition of Adaptive Capacity." *Global Environmental Change* 12 (1): 25–40.

Young, Kenneth R., and Jennifer K. Lipton. 2006. "Adaptive Governance and Climate Change in the Tropical Highlands of Western South America." *Climatic Change* 78: 63–102.

CHAPTER 8

Climate Change for Agrarian Societies in Drylands: Implications and Future Pathways

Simon Anderson, John Morton, and Camilla Toulmin

Global warming will bring major changes to rainfall and temperatures throughout the world, affecting the viability of many rural and urban livelihoods. Among the regions likely to be most affected are the world's drylands, given their existing exposure to drought and crop failure, high levels of poverty, and weak government services. Predicted changes in rainfall volume and distribution will have major consequences for the natural environment—soils, water, and vegetation—on which people and their livestock depend. Shifts in biophysical productivity will affect the social and economic characteristics of these regions. If adaptation to climate change is to be effective, the interactions among these different elements must be understood to enable adequate, equitable, and timely responses.

This chapter outlines the main impacts of climate change on the livelihoods of dryland peoples around the world. It focuses particularly on Africa and on the challenges faced by pastoral societies that are dependent on livestock. It reviews the predictions from science regarding likely changes in temperature and rainfall, and assesses the consequences for the main components of dryland incomes and assets. It sets the discussion of impacts within the larger context of changes currently under way, because climate change is only one of several forces affecting the social, economic, and environmental threads that are woven into the fabric of dryland livelihood systems. A full assessment of climate change impacts needs to review both direct and indirect effects associated with the range of measures and responses that global warming will bring. The chapter finishes with discussion of potential policy changes that would better shape adaptation to the multiple threats and opportunities that climate change will produce.

Location, Extent, and Ecological Status of Drylands

Drylands[1] cover 41 percent of the Earth's land surface; and they are home to more than 2 billion people, making up 35 percent of the world's population (Safriel and Adeel 2005, p. 625). The largest and most populated dryland regions are in developing countries, especially in Africa (the Sahel, the Horn of Africa, and southern Africa) and Asia (large parts of India and Pakistan, the dry steppes of Central Asia, and the Taklamakan and Gobi deserts in China and Mongolia). There are significant dry regions in South America, such as the northeast Brazilian cerrado and Patagonia in Argentina. Drylands are characterized by limited water resources, with seasonal, scarce, and unreliable rainfall leading to high variability in the water available for humans and animals and the moisture for plant growth. High temperatures cause much of the rainfall to be lost in evaporation; and the intensity of tropical storms means that much of the rain runs off, eroding the soil. Dryland areas experience substantial differences in rainfall both within the year and between years. Rainfall also is highly variable over short geographic distances: one village may receive an abundant shower while its neighbor 10 kilometers away remains dry.

Low levels of rainfall and high variability have led to a diverse range of coping strategies. These strategies include storage of surplus in the forms of gold, granaries, and livestock gathered in times of plenty and sold when times are hard; and risk spreading through diversification of activities and assets. Thus, it is common for dryland farmers to cultivate crops of different cycle lengths. They hope that if one fails, the others will be brought to harvest. Equally, herds are made up of sheep, goats, cattle, and sometimes camels. Investment in family and social networks is a vital strategy that provides support in times of crisis; it is expressed in the contacts, gifts, cooperation, and exchanges that make up daily life.

Human Populations and Well-Being in the Drylands

The characteristics of dryland ecosystems affect the well-being of inhabitants in many ways (Safriel and Adeel 2005). Unless there is access to irrigation, the limited water availability usually constrains the kind of cultivation that is possible to the short rainy season. Food security is at risk when several bad years follow each other. Dryland people face a number of health challenges resulting from malnutrition, poor access to clean drinking water, and exposure to disease and such parasites as guinea worm. Many dryland areas are poorly served by health and educational infrastructure as a result of their distance from urban centers and their low population densities. Periodic drought provokes livelihood shocks and population movements as people seek out employment, resources, and other support elsewhere.

More than 90 percent of dryland inhabitants are found in developing countries, and half of the world's poor people live in drylands.[2] They have the highest population growth rates of the world's major ecosystems, the lowest levels of human well-being, the lowest per capita income, and the highest infant mortality rates. Taking a look at the United Nations Development Programme's Human Development Index illustrates the low levels of well-being in many of the main dryland nations, with countries such as Burkina Faso, Chad, Ethiopia, Mali, and Niger occupying positions at the bottom of the index table. Thus, dryland dwellers are among the poorest on the planet, in large part because of the limited potential of the resources on which they depend and their vulnerability to environmental shocks. Such vulnerability will be tested further by the impacts of climate change over the decades to come.

Observed Climate Trends and Impacts

Changes in the climate of the world's drylands have already been observed. The Intergovernmental Panel on Climate Change (IPCC) reports an increase in the extent of dryland areas affected by more intense and longer droughts (IPCC 2007); and for the period 1900–2005, rainfall has been declining in southern Africa, northwest India, northwest Mexico, and most markedly in the Sahel (Trenberth and Jones 2007). In the Sahel, warming plus reduced rainfall have reduced the length of the growing season, which means that many of the traditional varieties of millet are not able to complete their growing cycle.[3]

It is not clear whether specific drought events and drought cycles are attributable to global warming. For example, the West African Sahel has experienced multidecadal periods of wetter and drier weather, interspersed with periodic harsh drought events (figure 8.1). But it is likely that global warming will exacerbate such droughts and other natural extremes. The current variable period most probably results from a combination of factors, such as sea-surface temperature rise, land degradation in the Sahel, and burning of tropical forest in coastal West Africa.

There is evidence that Africa is warming faster than the global average (Boko et al. 2007). Southern and western Africa have seen an increase in the number of warm spells and a decrease in the number of extremely cold days. In East Africa, by contrast, temperatures have fallen close to the coasts and major inland lakes (Boko et al. 2007). Overall, arid regions are expected to undergo significant changes as a result of global warming; but there is considerable variability and uncertainty in these estimates, depending on the different assumptions made. The complex interrelationships between rainfall changes; feedback mechanisms linking land surface, vegetation, and climate; and the effects of higher carbon dioxide levels on vegetative growth present particular challenges for modeling. A synthesis of projections for different dryland regions over the period 2080–99 is shown in table 8.1.

The table shows the marked upward trend expected in temperature, which will lead to higher levels of evaporation from soil, crops, and water bodies; and which will add stresses to human and animal health. It also shows that southern Africa, the Sahara, North Africa, and central Asia are projected to receive less rain and to experience more seasons and years

Figure 8.1. Variability in West African Rainfall, 1941–2001

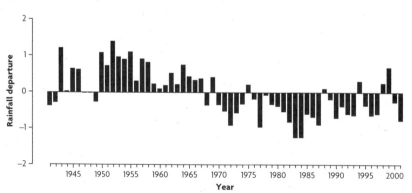

Source: Anuforom 2009.
Note. Time-series of average normalized April–October rainfall departure for 20 stations in the West African Sudano-Sahel zone (11-18°N and west of 10°E)

Table 8.1. Climate Change in Dryland Regions, Comparing Current Climate with Projection for 2080–99

Region	Median Projected Temperature Increase (°C)[a]	Median Projected Precipitation Increase (%)[a]	Agreement on Precipitation among Models[b]	Projected Frequency of Extremely Warm Years (%)[c]	Projected Frequency of Extremely Wet Years (%)	Projected Frequency of Extremely Dry Years (%)
West Africa	3.3	+2	Not strong	100	22	n.a.
East Africa	3.2	+7	Strong for increase in DJF, MAM, SON	100	30	1
Southern Africa	3.4	–4	Strong for decrease in JJA, SON	100	4	13
Sahara	3.6	–6	Strong for decrease in DJF, MAM	100	n.a.	[d]
Southern Europe and the Mediterranean	3.5	–12	Strong for decrease in all seasons	100	n.a.	46
Central Asia	3.7	–3	Strong for decrease in MAM, JJA	100	n.a.	12
Southern Asia	3.3	+11	Strong for increase in JJA, SON	100	39	3

Source: Adapted from Christensen et al. (2007), table 11.1.

Note: n.a. = not applicable; values are not shown in the original where fewer than 14 out of the 21 models agree on an increase or decrease in the extremes. JFMAMJJASOND = months of the year. It should be noted that the regions for which projections are given contain significant areas that are not drylands. The three analytical regions of Central and South America clearly are, in the majority, nondryland; and they are not included here. Reference should be made to the considerable amount of technical detail on, and qualifications to, the original table in Christensen et al. (2007).

a. The original source disaggregates median response by the four quarters of the year; figures in this column are the annual averages given in the original.

b. Agreement is "strong" (current authors' terminology) when the 25th percentile and the 75th percentile of the distribution of models were of the same sign.

c. Essentially, these are years warmer than the warmest year between 1980 and 1999; similar definitions apply for wet and dry years. The original source further presents projections of warm/wet/dry seasons.

d. No aggregate figure is given for years, but the frequency of dry DJF and MAM seasons is significant.

203

considered extremely dry than during the period 1980–99. East Africa and southern Asia are expected to receive higher levels of rain and a greater number of seasons and years that would be considered extremely wet, relative to 1980–99. The West African region presents considerable uncertainty, with significant disagreement among the different climate models regarding future trends in rainfall. Some models suggest a modest increase in rainfall for the Sahel, with little change on the Guinean coast; others predict much drier conditions.

Dryland Agrarian Societies

The physical geography of drylands imposes constraints on dryland livelihoods and associated social institutions. Political and economic constraints on dryland livelihoods also can be significant. Specific livelihood strategies respond to the biophysical, political, and economic constraints in various ways, which can be described under two broad themes: *coping and adaptation strategies* and *flexibility of institutions*.

Geographic Characteristics and Livelihood Strategies

Low and uncertain rainfall influences the productive base of dryland societies. Food production is dominated by maize, sorghum, and millet; with wheat in central Asia and the countries of West Asia and North Africa, combined with various legumes and root crops. Cash crops, though limited, include cereals, cotton, groundnuts, and soybeans in some areas. Livestock are extremely important for livelihoods, especially in drier parts where production systems usually are based on some form of pastoralism, which involves seasonal mobility of herds and people between grazing areas. Elsewhere, mixed-crop livestock systems prevail, with varying levels of interaction between crops and livestock.[4] Animal manure is often used as fertilizer and crop residues or cultivated forages serve as animal feed. Many farming systems rely on animal draught power; and livestock provide households with means to save, as well as a source of hides, meat, and milk. Cropping patterns rely on combining a range of different crops of different cycle lengths to reduce vulnerability to drought. Similarly, most farmers and herders keep a range of different livestock species, thereby reducing their vulnerability to risk. Rural dryland dwellers also rely on many other sources of income, such as collection of wild produce (leaves, medicines, game, woodfuel, and fish) and a wide range of off-farm activities (Toulmin et al. 2000).

Dryland livelihoods face a dynamic and ever-changing context linked to environmental variables, as well as to social, economic, and political changes. There are many adverse environmental trends—only some of which are related to climate change—that affect people's livelihoods, as set out by the Millennium Ecosystem Assessment (MA) (Safriel and Adeel 2005). For example, existing water shortages are projected to increase over time because of population increase; land cover change; increased competition between irrigated cropping, hydropower, and domestic needs; and global climate change. The conversion of rangeland areas to farmland is leading to a significant decrease in overall plant productivity and to restrictions on the pattern of mobile animal herding that had enabled people to thrive in these areas. The MA reckons that some 10–20 percent of the world's drylands suffer from one or more forms of land degradation, including erosion of soils through wind and water, mining of soil nutrients, and loss of plant cover (Safriel and Adeel 2005, p. 625). The MA also acknowledges that, in many areas, traditional and other management practices contribute to the sustainable use of the ecosystem and the services it provides; and, in some areas, dryland biodiversity remains relatively rich.

Coping, Vulnerability, and Adaptation

Dryland societies are characterized by an adaptive orientation and a readiness to adopt flexible strategies that reduce vulnerability ex ante to climate shocks and other forms of environmental risk.[5] Adaptations to climate variability usually involve practices such as pursuing a range of income-earning activities. Diversification can be achieved at individual, household, and community levels; and usually involves a combination of all three. Different activities are pursued according to season, in response to bad years, and sequentially so that people shift among cropping, various forms of mobile and sedentary livestock production, use of wild resources, off-farm employment, and out-migration. Within dryland cropping systems, adaptation strategies include use of multiple crops, livestock species, breeds, and varieties; intensified use of labor at critical times of the year (Mortimore and Adams 2001); on-farm storage of food and feed (Swearingen and Bencherifa 2000); and close management of soil and water resources. For pastoralists, mobility acts as a means to adapt to changing patterns of rainfall within and between years, whereas a combination of different livestock species—each herded in different directions—provides greater protection from drought. Rangeland tenure systems make it possible to move between grazing areas, through a range of secondary rights and reciprocal exchanges; and the buildup of large herds is a means to insure against drought (Morton 2006).

"Coping strategies" seek to manage the impacts of shocks ex post, and they include out-migration and recourse to wage labor, use of wild resources, exceptional pastoral migrations to drought refuges or rarely used rangelands, and intracommunity sharing of food and livestock. In extreme conditions, people must seek recourse to disaster relief services and even begging.[6] Coping with crisis and vulnerability is not free of cost, and neither is preparedness. For example, during the severe Sahelian droughts in the 1970s and 1980s, many farmers and herders moved southward, with several million people from Burkina Faso and Mali seeking land in the forests of Côte d'Ivoire. Such a large flow of incoming migrants and their success in developing large and productive plantations of coffee and cocoa put great pressure on land availability and generated considerable tension between incomers and local indigenous populations who felt they had lost control over their lands and heritage (Chauveau 2000). Much of the ongoing crisis in Côte d'Ivoire can be understood in terms of the competing claims over land among different groups within the country and their association with migrants from neighboring states.

Coping strategies may become adaptive strategies when people are forced to use them over a run of bad years and across seasons rather than just at the worst time of the year. Although not all coping strategies are erosive, they also can erode household assets, with the risk of households sinking into permanent destitution and degrading the environment (Davies 1993). Coping and adaptation are interlinked so that the way households cope with crises either may enhance or may constrain their future coping strategies, as well as their possibilities to adapt in the longer term. For example, if drought losses force households to sell off productive assets (such as work oxen) or send their productive labor to work for others, the households will find it increasingly difficult to restore their former levels of activity and income.

Eriksen, Brown, and Kelly (2005) have assessed how smallholder farmers in Kenya and Tanzania cope with drought and how different factors shape households' means of coping over time. They find that households in which an individual is able to specialize in one favored activity (such as employment or charcoal burning) were often less vulnerable than households where members are engaged in many diverse activities. Some groups, especially women, were excluded from certain more profitable activities, and had to rely on collecting wild fruit—an endeavor that yielded only marginal returns.

Out-migration by individuals or entire households is being used increasingly to cope with climate-related stresses, generating significantly different impacts on home and receiving areas. For example, in southern Mali, there has been substantial out-migration to neighboring Côte d'Ivoire by large numbers of family members over the last 30 years. This migration has led to substantial remittance flows back to the family home in Mali; and those flows then can be used for buying food and farm equipment as well as investing in shops and buying land.

Institutional Flexibility

Variability and risk in drylands have helped shape institutions[7] that allow diverse and complex patterns of resource use, such as seasonal access to different resources and interannual movement of people and animals.[8] Many such systems are traditionally based on a mix of farmland held at the individual and household levels, and wider areas of grazing and woodland managed under collective ownership at the lineage or village levels. In many cases, however, governments have asserted the underlying rights to land, leaving farmers and herders in a legally precarious position with their rights of management, occupancy, and use subject to the discretion of local officials.

Rangeland tenure in pastoral areas often is referred to as "common property," and occasionally as "open-access." Neither term does justice to the complexity and flexibility of rights in different resources (for example, grazing en passage, prolonged grazing, access to water, rights to cut trees and bushes for fodder, browsing) asserted, extended, and negotiated by different communities and collectives at different scales (Behnke 1994; Mendes 1988). Regarding tenure of cropped land, allocation of usufruct by traditional authorities remains very common, especially in Africa. There is a whole range of institutions (loans, cash rental, and sharecropping) that allocate secondary or "derived" rights in land in ways that may be crucial for spreading or managing risk (Lavigne Delville et al. 2001). Other examples of institutions include traditional patterns of loaning livestock and sharing their produce (Toulmin 1983), rights to use wells of different kinds, seasonal use of fallows and crop residues, collective labor parties, and networks for information (Morton 1988).

Institutional adaptations to uncertain dryland systems include patterns of domestic organization. For example, patterns of out-marriage mean that, in times of drought, families can call for help from people in other villages and areas that may not have suffered the same impacts. Equally, the persistence of large domestic social groups in the West African Sahel very likely is the result of the groups' great value as a means to increase people's resilience to multiple sources of risk (Toulmin 1992; Lewis 1979). Some households among the Bambara of central Mali may comprise 80 to 100 people, living together and working a common field (Toulmin 1992). Such large domestic groups enable people to thrive and prosper in circumstances that often are difficult. Being part of a large household provides great benefits because this group is better able to generate a surplus for reinvestment in farming and livestock production, to diversify activity, and to iron out fluctuations of adult-to-child dependency ratios because the many nuclear family units are operating together at different stages in their cycles.

Institutions such as tenure and social organization are fundamental in enabling adaptation and coping strategies by determining different people's access to various types of assets. Institutions tend to structure assets differentially for men and women, for groups defined by their landholder or outsider status or by age and lineage, and for castes and occupational groups. Women will have access to a different set of options for coping than will men. Market relations clearly favor the adaptation and coping of those who can mobilize salable assets, produce, savings or labor. Markets themselves are also embedded in local social relationships, allowing some people better access to flexible credit or other services. The existence and characteristics of rural organizations—such as elected local governments, membership organizations, cooperatives, service organizations, and private businesses—are important in structuring the range of adaptation options available. They can provide access to information, technology, financial capital, and leadership; but, as always, some groups are better able than others to tap into these resources.

Although there are clear links between the riskiness of dryland environments and patterns of diversification, mobility, and reciprocity, it cannot be assumed that the social institutions that have evolved in these contexts are suited to coping with major and unpredictable change. Neither is there any guarantee that adaptation pathways facilitated by these institutions are optimal at all times, nor that institutions are flexible enough to cope with climate change.

Economic and Political Context

Remoteness and low population density can exacerbate problems of political and economic marginalization. Where developing countries include both dryland and higher-potential areas, capital cities and centers of economic power usually are outside and at some distance from dryland areas (for example, Brazil, China, Ethiopia, Kenya, and the coastal states of West Africa). Dryland areas often are at the geographic margins of the state, lying close to national frontiers. Added to this is the fact that dryland population densities generally are low, leading to high per capita costs of providing infrastructure and services.

Pastoral peoples tend to suffer from multiple forms of marginalization (Markakis 2004; Lesorogol 1998): environmental, as a result of living in the lowest-potential areas; economic, because of poor market access and a low share of public expenditure; sociocultural, as a consequence of misunderstandings and prejudices; and political, as a result of all the above. Because pastoral groups often live close to and frequently cross international borders, they are seen as having only weak adherence to the nation and contributing little to the national economy. When they become involved in conflict over water, grazing, and raiding of herds, the international, national, and local causes become intermeshed. The current "global war on terror" has focused attention on the ungoverned frontier lands between many states, areas often occupied by nomadic groups. Thus, there is now the added threat to their livelihoods from military maneuvers and controls on their mobility.

Participation in markets can provide increased income, but often this comes at the cost of greater risk and dependency. Some countries depend greatly on exports of live animals and thus are very vulnerable to international veterinary regulations and "veterinary politics." In southern Africa, this takes the form of high costs to maintain access to European Union meat markets; and livestock producers from the Horn of Africa are exposed to the threat of import bans placed on animals by the Persian Gulf states. These bans have major consequences for the incomes not only of herders, but also of the many other people engaged in the livestock supply chain—from traders and brokers to transporters and butchers.

Poor service provision for health, education, and water helps account for the low level of human development in many dryland regions. Services to agriculture also are poor in many regions (unless bound up with produce markets, as in the Sahel cotton zones). Privatization or liberalization in the provision of these services has been followed to varying extents, as state systems have withered from a lack of resources. In some cases, such as in the Horn of Africa, the way forward for animal health services has been through the slow buildup of community-based provision, thanks to painstaking work by nongovernmental organizations (NGOs) and donor-funded bodies.

Governments often can be very distant from and unaware of the rationale for flexibility associated with dryland institutions, and then may undermine or disrupt ways of managing, using, and governing these resources. The lack of recognition of traditional land tenure and collective resource management (particularly among pastoralists) by so many governments is a classic case of such neglect that damages the productive capacity of such systems. Some progress is being made in parts of the Sahel, where the governments of Burkina Faso, Guinea, Mali, and Niger have agreed on new legal codes for the management of grazing lands that endorse the need for mobility and provide formal recognition of pastoral groups to manage certain resources.

Given the importance of mobility between different ecoregions, climate adaptation could be furthered by greater regional integration between neighboring countries. This integration would enable smaller, more vulnerable countries to share risks, coinvest in adaptive innovations, and allow for greater planned and spontaneous movement between their territories in response to climate vagaries (IIED 2008). Currently, such mobility is hampered by government attempts to control the movements of livestock and people rather than to recognize the need for this interstate reciprocity.

Impacts of Climate Change

The impacts that climate change will bring can be categorized into a range of direct impacts—including changes in temperature, a shift in rainfall patterns and distribution, and sea-level rise—that operate both individually and in combination at different scales, as outlined in table 8.2. And a range of indirect impacts linked to the policies adopted to address climate change, such as renewable fuel targets, will generate a set of impacts on land values and returns to different forms of crop production, among other things.

Direct Impacts

Table 8.2 presents a framework to integrate different sorts of impact. The most obvious impacts, and the most important in the medium term, are those identified in the table as "extreme events," especially increased risk of droughts and floods. Trends seen over the last few decades of more

Table 8.2. Direct Impacts of Climate Change on Dryland Systems

Impact	Field/Organism	Landscape/Environment	Human Health	Infrastructure and Nonagricultural Employment
Extreme events	• Increased droughts • Increased chances of failed growing seasons • Increased floods • Increased outbreaks of rainfall-related animal disease, such as Rift Valley Fever • Increased snow-related disasters in colder areas	Increased soil erosion, resulting from floods	Increase in waterborne diseases during floods	Destruction of infrastructure in drylands and adjacent urban areas through floods and sea-level rise
Increased variability in rainfall and temperature	Specific effects on crop development, such as shorter growing seasons	Increased soil erosion, resulting from heavier erosiveness of rainfall	• Greater risks of food insecurity • Shifts in climate-related disease prevalence, such as Dengue Fever	Damage to roads, resulting from heavier erosiveness of rainfall
Shift in average temperature and rainfall	• Higher crop water demands, resulting from warming • Decreased rainfall in some areas • Shorter growing periods • Direct carbon dioxide fertilization • Declining yields, especially maize • Increased heat stress to cattle	• Remobilization of sand dunes • Increased encroachment of woody species on rangeland • Decreased river flows, less water available for irrigation and hydropower • Shifts in boundary of tsetse fly/trypanosomiasis	Shifts in malaria and meningitis	Decreased tourism revenue and employment in tropical countries, resulting from warming and water shortages

Source: Morton 2007.

frequent and severe droughts in many dryland regions are increasingly being linked to global climate change. Apart from the complex case of the Sahel, this has been most marked in the Horn of Africa; but unprecedented multiyear droughts have also struck North Africa and western Asia. Floods have become more frequent in some dryland regions, such as the Horn of Africa and Mozambique, largely because of intense rainfall in upstream areas. For example, loss of vegetation and the consequent reduction in the water-holding ability of soils in the higher Zambezi River catchment have been cited as contributory factors behind the heavy flooding in Mozambique in 2000 and 2007.

The IPCC's fourth assessment projects a likely increase in extremely wet and extremely dry years in most dryland regions (table 8.1), a finding that is echoed by Burke, Brown, and Christidis (2006) who predict that the proportion of the world's land surface at risk of extreme drought will increase markedly over this century. More specific risks of extreme events projected for the future include an increased number of outbreaks of Rift Valley Fever, which is associated with heavy flooding in the Horn of Africa and which has devastating consequences for livestock and the people depending on them for livelihoods (Baylis and Githeko 2006).[9] There are also predictions of a rise in extreme cold or snow events, known in Mongolia as *dzud* (Batima 2006). The impact of extreme events also will be felt, for example, through increased soil erosion, a greater exposure to waterborne diseases, and the damage or destruction of infrastructure by floods (loss of bridges and buildings and silting up of dams).

The impact of a change in average climate parameters is likely to be felt over longer time scales. The principal impact on crops will result from a combination of higher water demands linked to rising temperatures and decreased rainfall causing greater plant stress in most dryland regions. These negative effects may be offset to some extent by the positive effect that higher atmospheric levels of carbon dioxide can have on increasing crop growth; but recent research has tended to play down that factor (Easterling et al. 2007). Changes to rainfall and temperature will also affect insect pollinators (such as bees) and the incidence of crop pests and diseases (such as wheat rust and striga).

A synthesis of recent modeling studies for maize, wheat, and rice has been carried out by the IPCC to assess how rising temperatures and changes to rainfall will affect yields (Easterling et al. 2007). For maize, the most important of the three crops in dryland regions, the models show a clear downward trend in yields. Jones and Thornton (2003) show that aggregate yields of maize in Africa and Latin America (largely but by no means exclusively cultivated in dryland regions) are likely to decrease by 10 percent by 2055 (p. 51). These results hide great variability between regions and give cause for concern, especially in some areas of subsistence agriculture. Another approach, which models average length of the growing period, shows that large areas of southern Africa, the Maghreb, the Sahel countries, and Sudan will suffer a 20 percent loss in length of the growing period by 2050,[10] bringing a significant increase in the number of failed rainy seasons, especially in the Sahel (Thornton et al. 2006, p. 44). Drucker et al. (2008) review the current literature on how climate change impacts will affect different livestock species and breeds in six regions of the world. They suggest that climate change can be expected to affect livestock productivity directly, making breeds with greater heat or cold tolerance more attractive; and indirectly through its effect on the availability of feed and fodder and on the presence of diseases and parasites. However, such estimates made at a global or regional level hide complex spatial patterns of change, with impacts dependent on circumstance and on the response from producers, government, and other key actors.

There also are important climate change consequences that operate at the larger landscape level—for instance, as a result of changes in river flow. Many dryland regions rely on major rivers bringing water from afar. Examples include the River Nile in Africa, which is fed by rainfall in the highlands of Ethiopia and Uganda; and the Indus River, fed by the Himalayan snow, on which rely the large irrigation systems of the Indo-Gangetic plain in India and Pakistan. Rising temperatures and shifts in rainfall and snow will change the volume and timing of water flow in these major rivers in ways that may greatly damage the continued capacity of highly productive systems of irrigation to work as they do now. Barnett, Adam, and Lettenmaier (2005) highlight the risks of a significant fall in usable water in such systems as a result of snowpack melt, jeopardizing the harvest on which many millions of people in southern Asia depend.

The IPCC (Easterling et al. 2007) points to ways in which climate change will affect the structure and quality of rangeland vegetation. For example, there is likely to be greater encroachment by woody species on many rangeland areas, as a result of higher carbon dioxide levels in dry parts of North America (Morgan et al. 2007). Shifts in the distribution of the tsetse fly, which likes denser bush and damper conditions, will generate impacts on agriculture and on human health. But those effects may be relatively minor, given the potential for control of tsetse flies through insecticides and bush clearance (Baylis and Githeko 2006). Finally, there are impacts on nonagricultural livelihoods, such as a projected decline in tropical tourism because of higher temperatures, a fall in water supplies, and greater stress in receiving areas (Wilbanks et al. 2007). Tourism is a major employer in many developing countries, including dryland regions that host major game parks.

Indirect Impacts

People's production systems and livelihoods also will experience indirect impacts from changes generated by global warming, as well as through the measures they take to cope, adapt, and respond. Policies adopted to address climate change may produce a range of unintended consequences, such as incentives to greatly expand the area of land under biofuel cultivation, and changes in the pricing of carbon and its impacts on the value of tropical forestland. These unintended consequences will feed through in the form of local changes in land use and land tenure (see SOS Sahel 2008); and, at national and global levels, through changes in food prices because of large-scale shifts in agricultural production from food to biofuel cultivation (Mitchell 2008; Wiggins et al. 2008). A simplified illustration of these interactions is presented in figure 8.2.

Aggregate Impacts

Given the number of and complex interactions among different variables, it is helpful to provide a broader synthesis, as offered by Thornton et al. (2006). They combine modeling the length of the growing period with analysis of socioeconomic vulnerability by farming system to identify certain "hot spots" in Africa that are likely to be most vulnerable. These spots are the semiarid, rain-fed crop-livestock systems of the Sahel, the arid and semiarid pastoral grazing systems of East Africa, and the crop-livestock and highland perennial crop systems found in the Great Lakes region of central Africa.

Both more frequent drought and rising levels of risk will increase fluctuations in the production of crops and livestock, and may lower the level of investment made by farmers and livestock producers. This will take the form of reduced investment of labor, a decreased tendency to adopt new technology, and poor development of markets for those livestock and crop commodities suffering unpredictable supply. Heightened risks to cropping could lead to an increase in more extensive livestock production systems and greater diversification of income-generating activities, such

Figure 8.2. Climate and Nonclimate Impacts on Dryland Livelihoods

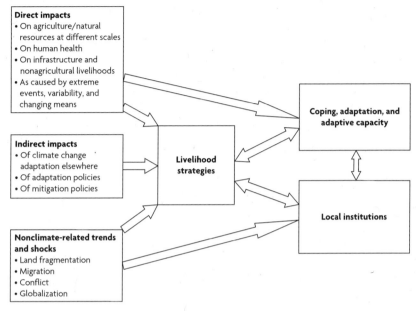

Source: Authors' illustration.

as migration and off-farm employment. However, the policy environment needed to support such shifts in livelihoods is often lacking. Governments frequently seek to curb the mobility on which extensive livestock production depends, to constrain labor migration, and to offer little help with the secure transfer of remittance income from migrants to their families back home.

Dryland dwellers have a repertoire of adaptive strategies by which they have tried to cope with climate variability in the past. These strategies tend to combine the storage of assets during times of plenty and the diversification of income-generating activities, with investment in a geographically spread network of support. The drought years of the 1970s and 1980s in the Horn of Africa and beyond provide many examples of how different people were able to cope with losses to harvests and herds. These examples include early migration to wetter areas; sales of livestock before they had weakened and prices had slumped; sales of jewelry; and divisions of households, with women and children seeking food and shelter among their kin or in feeding camps while men set off to seek work in town or beyond.

But many coping strategies—especially those accessible to the poor and people made more vulnerable by other factors (such as loss of land)—will erode household assets and environmental sustainability, thereby making it almost impossible to regain an independent household existence when conditions improve. Examples include herders being forced to sell off breeding females, on which the viability of their future herd depends; farmers putting themselves and their sons out to work for others instead of preparing their own fields; and villagers cutting forest resources for sale of firewood and charcoal at the expense of longer-term balance in resource flows between croplands, grazing areas, and woodlands.

In terms of institutions, the impacts of climate change are hard to gauge. The flexibility associated with institutions within dryland societies is important for adaptation to both climate variability and climate change. Climate change per se is unlikely to weaken that flexibility, but policy trends over recent decades have imposed further stresses on people's capacity to cope. In much of Africa, for example, governments have been seeking to "modernize" their agricultural sector by settling pastoral herders, establishing ranches, and developing large-scale commercial agriculture in place of smallholder farms. Past experience shows that such versions of "modernization" are often far more vulnerable to environmental and economic shock than are the more diverse smallholder systems they supplant. It remains to be seen whether newer policy trends for increased local self-governance, such as decentralization and increased recognition of community resource management, can strengthen adaptive capacity at the local level. There are also reasons to fear that governments may use anxiety about climate change and other forms of environmental degradation to increase bureaucratic and political control over key dryland resources (Toulmin 2009; Leach and Mearns 1996).

Policy Trends Intertwined with Climate Change Impacts

Over the past 20 years, many parts of the developing world have been adopting new policy measures, often at the instigation of donor agencies. For example, following structural adjustment to cope with budget deficits in the 1980s, there has been a general trend for *state withdrawal from markets and services* across Africa. In central Asia and Mongolia, a rapid decollectivization was associated with the collapse of communism, and there was the need to establish new structures for management of land and grazing and to assure the delivery of services. Other dryland regions, such as those in North Africa, western Asia, and India, had less state involvement in the rural economy to begin with, and thus both the degree and the impact of withdrawal have been less marked.

In dryland Africa, the pattern of withdrawal has been mixed, depending on region and sector. In Kenya in the early 1980s, the role played by government as buyer of last resort for livestock (through the Livestock Marketing Division and the Kenya Meat Commission) was ended. Whatever the arguments about the unsustainability of government purchase, this shift is still regretted by Kenyan pastoralists. In Ethiopia, by contrast, the effects of the early-1990s liberalization of both meat and livestock exports have been broadly positive, even though much remains to be done to facilitate greater competition and efficiency in private sector activity and marketing chains. Withdrawal from input provision and extension has been more universal, followed by slow steps to provide alternative systems for delivery of veterinary services and crop inputs, with increased reliance on private suppliers, farmer organizations, and NGOs.

A rather different pattern is evident for cotton production in the Sahelian countries, where parastatal companies partly owned by both national governments and the French government have long controlled a system of crop purchase, interlocked with credit, inputs, information, and extension work. Until recently, countries have resisted the pressure to liberalize, with a range of tactics being adopted in the privatization debate.

Parallel with economic liberalization, governments (particularly in Africa, but also in northeast Brazil) have pursued policies of *decentralization* aimed at establishing elected local government structures. In principle, decentralization of governance encourages local decision making and the potential space for local adaptation and building of local capacity; but there also is the risk that decentralization will be captured by local elites, and will favor settled people at the expense of pastoralists and other mobile people. Given the very low tax base, it also is likely that local governments will be unable to deliver expected services because of high transport and transaction costs in extensive areas with low population density. Early findings from decentralization in Mali and Niger suggest that many of the same political battles that exist at national level are also played out in local government, while the very limited resources available at the local level receive very little supplementation from the central government. At the same time, central government has been unwilling to relinquish control over valuable resources that might generate revenue for local authorities.

Land tenure policies are undergoing major changes and have taken many forms, including the registration of rights at the household and community levels. In many dryland areas of sub-Saharan Africa, the large majority of land has no formal documentation of rights, with most rural land users gaining access through a range of informal mechanisms (such as sharecropping, loans of land, and seasonal rights negotiated with local chiefs). In many countries, governments assert fundamental ownership of land in a way that enables those governments to reallocate areas of land if they so wish, without having to pay significant compensation to those occupying the land. Grazing lands are particularly subject to such seizures, with governments often claiming the land is not being put to productive use and hence can be reallocated. However, some countries (such as Kenya), are now progressing slowly toward a clearer recognition of collective grazing rights.

Some national governments responsible for dryland regions have long taken a role in "managing drought" (Morton 2002) through food relief, public works programs to create temporary employment, and subsidized feed distribution for livestock (Oram 1998). Such strategies have been followed for decades by India and some countries in North Africa and western Asia. In the 2001 drought, the Government of India distributed 500,000 tonnes of grain through Food for Work and 170,000 tonnes as free animal feed, and provided free rail transport for relief goods, and cash grants to state governments (UNDP 2001, p. 51). Governments of poorer developing countries usually rely on emergency assistance from international donors, often delivered long after protracted assessment, negotiation, and logistics.

The African droughts of the mid-1980s brought about new forms of drought management, with an emphasis on linking early-warning systems, contingency planning, and mitigation measures to support livelihoods. In countries such as Ethiopia, these safety nets have acquired permanent status, so that even in a normal year, more than 8 million people receive a cash grant for buying grains and meeting other needs (DFID 2008). New technologies—particularly, remote sensing of vegetation conditions—have been deployed to give early warning of drought so that national governments and donors can take appropriate actions, including prompt delivery of food relief.

A powerful model of district-level contingency planning was developed after early experience in Turkana District, Kenya (Swift 2001). Regular on-the-ground data collection is designed and implemented so that results can be codified into "warning stages," moving from normal to alert, then alarm, followed by emergency, and subsequent recovery. At the same time, contingency plans are made at district or similar levels for relief, mitigation, and rehabilitation actions that can be triggered at specific warning stages. These early-warning systems then can be combined with measures to buffer the adverse effects of drought by providing employment in large-scale public works. The Relief and Rehabilitation Commission of Ethiopia has maintained such works for decades, building roads and constructing many kilometers of stone terraces on farmland and hillsides. Equally, Indian state governments long have resorted to relief works in times of drought and hunger.

In pastoral and livestock-based livelihood systems, a number of innovative pre- and postdrought strategies have been pilot-tested to maintain livestock numbers and improve livestock-keeper purchasing power during drought. They include rapid and early offtake of livestock; destocking, emergency slaughter, and meat distribution; supplementary feeding for livestock; water provision for livestock; veterinary care; and livestock shelters (LEGS 2008). As well as reducing animal numbers, herders can receive help by negotiating special access to protected areas and, through peace building, can reach pastures normally closed off because of endemic conflict.

Whereas drought management is being mainstreamed in work funded by international donors,[11] greater learning from customary and indigenous drought-coping strategies would be valuable. Newer approaches, such as the use of index-based insurance,[12] are being tested. However, drought management cannot be divorced from a whole range of longer-term policies (most notably on land tenure, resource management, and access to markets), which can strengthen or weaken resilience to drought. In the context of climate change, such links are ever-more important.

Political Processes and Adaptation

Poor policy limits people's capacity to adapt to the changes being brought by global warming. There persist among many governments the beliefs that local people are a problem and that their knowledge and ways of life are outdated. A set of policy narratives has remained very strong, and continues to influence how governments intervene in these areas (Swift 1996). For example, such initiatives emphasize the need to stay below the "carrying capacity," stressing the problem of "overgrazing" and "the trag-edy of the commons." Research in the past 15–20 years has shown these concepts to be unhelpful. Many dryland grazing systems are well managed by local pastoral systems, given the highly uncertain patterns of rainfall (rather than "overgrazing") that are the primary cause of shifts in vegeta-tion availability.

Even where progressive, well-informed policies are designed, their implementation depends on many variables. Administrative structures are frequently poorly staffed and weak, so that policy implementation in the drylands, in practice, is often ineffective, inefficient, or partial. One advantage for many dryland peoples remains the fact that government is relatively absent, so the potential damage from too active and interested a presence (with its bureaucratic exactions and interference) is limited.

The analysis of dryland *governance* requires a focus on how dryland dwellers are represented in decision-making and implementation processes. In her study of Ethiopian pastoralist representation, Lister (2004) warns of the dangers of making judgments about the representative quality of particular institutions or individuals. She analyzes a number of "processes mediating between citizen interests and policy outcomes... the functioning of the federal parliament, the functioning of regional and sub-regional sys-tems of government, and the interaction between formal and 'traditional' or 'customary' institutions" (p. 28), which add up to a far-from-perfect picture of pastoralist representation. However, she does note that the pat-tern of representation is shifting gradually in the right direction, with ordi-nary people having heightened expectations of what they should hope for from their governments. Morton, Livingstone, and Mussa (2007) examine the groups that have sprung up to represent pastoralists in the parliaments of Ethiopia, Kenya, and Uganda. They demonstrate a certain potential for those groups to promote pro-poor development and to offer opportunities for proposals for donors and NGOs to engage with them.

There are many other strategies for increasing dryland dwellers' voice and influence on structures of power. These strategies include support for civil society groups that can build links to formal government and pol-icy makers and can act above the community level. Dryland dwellers can act politically through producer organizations focusing on specific com-modities or economic sectors, especially where these are federated at the regional or national level. The national livestock-producer federations of the Sahelian countries are good examples. Some of the new hybrid struc-tures emerging at the local level may be effective in representing dryland dwellers; and, in some countries, new experiments are being tried—for example, the Council of Amakari (elders) in Somali Region, Ethiopia. While decentralization is being rolled out as a mode of making and implementing policy, there needs to be a more active means to encourage poorer, more vulnerable groups (such as women) to participate in local decision making. The tendency for decentralization to assume a territorial model, with clearly defined boundaries, may discriminate against the participation of mobile peoples, such as pastoralists.

A number of countries offer interesting examples, like the representation of particularly vulnerable groups through appointed bureaucratic bodies (such as state-level welfare boards for particular castes, tribal peoples, and nomads in India). Improvements can also be made in the ways mass media deal with the problems faced by dryland dwellers, by providing better information and promoting new uses of communication technology (Internet, video, mobile phones, and FM radio) to overcome the constraints of distance and information that limit participation in policy and political processes.

Implications for Policy

As argued above, climate change and its multiple consequences will exert profound impacts on dryland areas in developing countries and on the people who live in those areas. There are many possible interventions for strengthening adaptation to climate change in dryland areas. Some of these interventions are immediate and relate to ensuring a strong presence and voice for those affected countries and communities during the period of negotiations for the successor to the Kyoto Protocol now under way. Other interventions concern the need to support adaptation in the immediate term, combined with longer-term investment in infrastructure for disaster risk reduction, clean energy, and better water management.[13]

Most poor countries in Africa and elsewhere have only recently started paying serious attention to the impacts of climate change and the need to support adaptation. Therefore, significant gaps existing in policy and practice merit investment in pilot-testing and lesson learning (including mainstreaming adaptation to climate change into national policy making, particularly at the sectoral level). Doing so would focus on disaster risk reduction from climatic disasters (such as floods and droughts), water resources management, climate-proofing agriculture, and coastal zone management. This process of mainstreaming must also pay attention to the large increase expected in urban populations; and to ways to enhance resilience and reduce vulnerability in towns and cities, given their vital role in local, national, and regional economic networks.

Experience from the West African Sahel, following the major droughts of the 1970s and 1980s, shows the benefits gained when neighboring countries share ideas and practical solutions for coping with drought and climate variability. Advancing regional cooperation presents a valuable opportunity for avoiding duplication and for scaling up successful initiatives (such as pilot-testing adaptation projects with poor and vulnerable communities in dryland areas; and building research capacity, knowledge sharing, and dissemination activities within regional and national institutions).

Biofuel production offers a potentially valuable source of revenue and energy for local needs as well as export with crops (such as jatropha) that are well suited to semiarid regions. However, the recent rapid expansion of biofuel production schemes highlights important trade-offs between ensuring energy supply and increasing pressures on scarce land and water resources. It also points out risks to poor people's land rights and access and to their food security. There is useful work to be done to build an evidence-based assessment of the conditions under which biofuel production generates positive or negative impacts on environmental, social, and economic dimensions.

Many African and other low-income countries have a very strong interest in ensuring rapid, ambitious, and effective cuts in greenhouse gas emissions, thereby maximizing the probability of limiting the global temperature rise to 2 degrees Celsius. Every delay in meeting tough targets for emission cuts will bring higher temperatures, increased damage, and greater costs and difficulties in adaptation—particularly for poor countries and vulnerable communities in the developing world. It is vital that these interests are able to engage with and shape the post-2012 agreement effectively, making sure it is responsive to their needs and priorities. Doing so requires work to ensure that the needs of the most vulnerable countries are brought to the fore. A mix of activities will be needed: strengthening negotiating capacity, tactics, and strategy; developing input from research, civil society, and community groups to provide effective grassroots voice to national positions; assessing costs of climate change for different countries and sectors; and testing practical options for adaptation. Also included will be programs of public information and education to generate public understanding of the challenges to be faced.

Most of the important direct impacts of climate change are associated with water, in one form or another—whether it be increased drought, heavier storms, flooding, or sea-level rise. Thus, for dryland regions, a strong focus on improved management of water at local, national, and regional levels will be key to building greater resilience to disasters, crop failures, and damage to infrastructure. This requires a combination of microinvestments at the field and village levels, with wider water catchment management and coordinated planning of major river basins—such as those of the Niger, Nile, and Zambezi.

Better access to good climate information in a form that is comprehensible and of value to local people and their organizations would allow for better planning and disaster preparedness. New communication technologies provide possible means for spreading information rapidly and widely in ways that could bring much innovation to the management of drought, floods, and other extreme events. There also may be untapped opportunities for dryland regions associated with carbon markets to achieve stronger synergies between adaptation and mitigation through the design of soil, pasture, and woodland management practices that can supply carbon sinks on a significant scale (Reij 2008).

Conclusion

Climate change presents us with most profound and difficult challenges, both in designing ambitious and robust targets for mitigation and in finding ways to support adaptation around the world (especially in those regions most vulnerable to adverse impacts). There is much we need to learn as we engage in the testing of adaptation in practice. We have very little evidence on how the costs and benefits of climate adaptation are distributed in reality and how greater equity can be brought into the distribution process. Over the next few years, we need to build an effective means for sharing experience with good and bad adaptation practices; learning about the best mix of community-based processes and district, national, and subregional actions; and implementing the right blend of short-, medium-, and longer-term adaptation options for investment and institutional support, such as development of insurance schemes.

It is said that you can tell the quality of a government and a society by how well they address the needs of their poorest citizens. In like manner, we will be judged as global citizens by how well we respond to the growing difficulties being faced by many of the poorest people in the world, whose livelihoods are being transformed in damaging ways by climate change.

Notes

1. Drylands are defined by the Millennium Ecosystem Assessment (MA), as "all terrestrial regions where water scarcity limits the production of crops, forage, wood and other ecosystem provisioning services" (MA 2005, p. 1).
2. Data in this section are taken from the MA.
3. For information on millet, see Ben Mohamed, van Duivenbooden, and Abdoussallam (2002); for information on groundnuts, see van Duivenbooden, Abdoussallam, and Ben Mohamed (2002).
4. See Scoones and Wolmer (2005) for an "unpicking" of this generalization.
5. "Adaptation" includes "adjustments, or changes in decision environments, which might ultimately enhance resilience or reduce vulnerability to observed or expected changes in climate" (Adger et al. 2007, p. 720); and it is linked closely to adaptive capacity: "the ability or potential of a system to respond successfully to climate variability and change" (p. 727).
6. For information about the Sahel, see Davies (1996); for information about pastoralists in the Horn of Africa, see Morton (2006).
7. Institutions can be formal and informal; and they can have political and economic functions, such as establishing and protecting property rights, facilitating transactions, and permitting economic cooperation and organization (Wiggins and Davis 2006). More broadly, they can regulate access to the various factors of production—land, labor, capital, and information (Scoones and Wolmer 2005).
8. Independent pastoral development researcher Saverio Krätli (personal communication) points out that pastoralist groups in Niger know how the same ecosystem components can be used by different groups of dryland dwellers in different ways. Thus, the "resource" is defined by use and not by its inherent characteristics and abundance.
9. Rift Valley Fever is dangerous to both livestock and humans. But in recent years it has become important in the Horn of Africa as the cause of economically devastating livestock trade bans imposed by the Persian Gulf states. The scientific basis for those bans has been questioned by international agencies and authorities in the exporting countries.
10. Using the U.K. Met Office's Hadley Centre Coupled Model, version 3 (HadCM3) climate model, and the IPCC's (2000) emissions scenario A1F1.
11. At an operational level, this is occurring through the Kenya Arid Lands Resource Management Program, the Ethiopia Pastoral Community Development Program, and the Mongolia Sustainable Livelihoods Program, all funded

by the World Bank; at the level of developing policies, it is occurring through the ALive initiative (ALive 2007) and the Livestock Emergency Guidelines and Standards project (LEGS 2008).

12. With index-based insurance, farmers or livestock keepers can be insured against area-based indexes (of rainfall, vegetation quality, or livestock mortality) passing certain thresholds, thus avoiding the moral hazard and adverse selection problems associated with agricultural insurance in developing countries. This approach has been applied in the Mongolia Index-Based Livestock Insurance Project, funded by the World Bank.

13. The Commission on Climate Change and Development, launched by the Swedish government, provides a valuable opportunity to highlight issues of adaptation and disaster risk reduction (Commission on Climate Change and Development 2009).

References

Adger, W. Neil, Shardul Agrawala, M. Monirul Qader Mirza, Cecilia Conde, Karen O'Brien, Juan Pulhin, Roger Pulwarty, Barry Smit, and Kiyoshi Takahashi. 2007. "Assessment of Adaptation Practices, Options, Constraints and Capacity." In *Climate Change 2007: Impacts, Adaptation and Vulnerability. Contribution of Working Group II to the Fourth Assessment Report of the Intergovernmental Panel on Climate Change,* ed. Martin L. Parry, Osvaldo F. Canziani, Jean Palutikof, Paul van der Linden, and Clair Hanson, 717–43. Cambridge U.K.: Cambridge University Press.

ALive (Partnership for Livestock Development, Poverty Alleviation, and Sustainable Growth). 2007. "Community-Based Drought Management for the Pastoral Livestock Sector in Sub-Saharan Africa." Policy Note, ALive, Washington, DC.

Anuforom, Anthony C. 2009. "Climate Change Impacts in Different Agro-Ecological Zones of West Africa—Humid Zone." Paper presented at the International Workshop on Adaptation to Climate Change in West African Agriculture, Ouagadougou, Burkina Faso, April 27–30.

Barnett, Tim P., Jennifer C. Adam, and Dennis P. Lettenmaier. 2005. "Potential Impacts of a Warming Climate on Water Availability in Snow-Dominated Regions." *Nature* 438: 303–09.

Batima, Punsalmaa. 2006. "Climate Change Vulnerability and Adaptation in the Livestock Sector of Mongolia." Assessments of Impacts and Adaptations to Climate Change Final Report, Project AS 06, International START Secretariat, Washington, DC.

Baylis, Matthew, and Andrew K. Githeko. 2006. "The Effects of Climate Change on Infectious Diseases of Animals." Review prepared for the Foresight Project,

"Infectious Diseases: Preparing for the Future," Office of Science and Innovation, London. http://www.foresight.gov.uk/Infectious%20Diseases/t7_3.pdf.

Behnke, Roy. 1994. "Natural Resource Management in Pastoral Africa." *Development Policy Review* 12 (1): 5–28.

Ben Mohamed, M., N. van Duivenbooden, and S. Abdoussallam. 2002. "Impact of Climate Change on Agricultural Production in the Sahel. Part 1: Methodological Approach and Case Study for Millet in Niger." *Climatic Change* 54 (3): 327–48.

Boko, Michel, Isabelle Niang, Anthony Nyong, Coleen Vogel, Andrew Githenko, Mahmoud Medany, Balgis Osman-Elasha, Ramadjita Tabo, and Pius Yanda. 2007. "Africa." In *Climate Change 2007: Impacts, Adaptation and Vulnerability. Contribution of Working Group II to the Fourth Assessment Report of the Intergovernmental Panel on Climate Change,* ed. Martin L. Parry, Osvaldo F. Canziani, Jean Palutikof, Paul van der Linden, and Clair Hanson, 433–67. Cambridge, U.K.: Cambridge University Press.

Burke, Eleanor J., Simon J. Brown, and Nikolaos Christidis. 2006. "Modeling the Recent Evolution of Global Drought and Projections for the Twenty-First Century with the Hadley Centre Climate Model." *Journal of Hydrometeorology* 7 (5): 1113–25.

Chauveau, Jean-Pierre. 2000. "The Land Question in Côte d'Ivoire: A Lesson in History." Drylands Programme Issue Paper 95, International Institute for Environment and Development, London.

Christensen, Jens Hesselbjerg, Bruce Hewitson, Aristita Busuioc, Anthony Chen, Xuejie Gao, Isaac Held, Richard Jones, Rupa Kumar Kolli, Won-Tae Kwon, René Laprise, Victor Magaña Rueda, Linda Mearns, Claudio Guillermo Menéndez, Jouni Räisänen, Annette Rinke, Abdoulaye Sarr, and Penny Whetton. 2007. "Regional Climate Projections." In *Climate Change 2007: The Physical Science Basis. Contribution of Working Group I to the Fourth Assessment Report of the Intergovernmental Panel on Climate Change,* ed. Susan Solomon, Dahe Qin, Martin Manning, Melinda Marquis, Kristen Averyt, Melinda M. B. Tignor, Henry LeRoy Miller Jr., and Zhenlin Chen, 847–940. Cambridge, U.K.: Cambridge University Press.

Commission on Climate Change and Development. 2009. *Closing the Gaps: Disaster Risk Reduction and Adaptation to Climate Change in Developing Countries.* Final Report of the Commission on Climate Change and Development. Stockholm, Sweden: Ministry of Foreign Affairs.

Davies, Susanna. 1993. "Are Coping Strategies a Cop Out?" *IDS Bulletin* 24 (4): 60–72.

———. 1996. *Adaptable Livelihoods: Coping with Food Insecurity in the Malian Sahel.* Basingstoke, U.K.: Palgrave Macmillan.

DFID (U.K. Department for International Development). 2008. "A Safety Net against Famine in Ethiopia." London: DFID. http://www.dfid.gov.uk/Media-Room/Case-Studies/2008/A-safety-net-against-famine-in-Ethiopia/.

Drucker, A. G., S. J. Hiemstra, N. Louwaars, J. K. Oldenbroek, and M. W. Tvedt. 2008. "Riding Out the Storm: Animal Genetic Resources Policy Options under Climate Change." Paper presented at the International Conference on Livestock and Global Climate Change, Hammamet, Tunisia, May 17–20.

Easterling, William, Pramod Aggarwal, Punsulmaa Batima, Keith Brander, Lin Erda, Mark Howden, Andrei Kirilenko, John Morton, Jean-François Soussana, Josef Schmidhuber, and Francesco Tubiello. 2007. "Food, Fibre and Forest Products." In *Climate Change 2007: Impacts, Adaptation and Vulnerability. Contribution of Working Group II to the Fourth Assessment Report of the Intergovernmental Panel on Climate Change,* ed. Martin L. Parry, Osvaldo F. Canziani, Jean Palutikof, Paul van der Linden, and Clair Hanson, 273–313. Cambridge, U.K.: Cambridge University Press.

Eriksen, Siri H., Katrina Brown, and P. Mitch Kelly. 2005. "The Dynamics of Vulnerability: Locating Coping Strategies in Kenya and Tanzania." *Geographical Journal* 171 (4): 287–305.

IIED (International Institute for Environment and Development). 2008. "Adaptation to Climate Change in Africa." Presentation to the Technical Working Group on Climate Change and Development in preparation for the Seventh Nordic-Africa Foreign Ministers Meeting, Copenhagen, Denmark, February 14.

IPCC (Intergovernmental Panel on Climate Change). 2000. "Summary for Policy Makers." In *Special Report on Emissions Scenarios,* ed. Nebojsa Nakicenovic and Rob Swart. Cambridge, U.K.: Cambridge University Press.

———. 2007. "Summary for Policy Makers." In *Climate Change 2007: The Physical Science Basis. Contribution of Working Group I to the Fourth Assessment Report of the Intergovernmental Panel on Climate Change,* ed. Susan Solomon, Dahe Qin, Martin Manning, Melinda Marquis, Kristen Averyt, Melinda M. B. Tignor, Henry LeRoy Miller Jr., and Zhenlin Chen. Cambridge, U.K.: Cambridge University Press.

Jones, Peter G., and Philip K. Thornton. 2003. "The Potential Impacts of Climate Change on Maize Production in Africa and Latin America in 2055." *Global Environmental Change* 13 (1): 51–59.

Lavigne Deville, Philippe, Camilla Toulmin, Jean-Philippe Colin, and Jean-Pierre Chauveau. 2001. *Negotiating Access to Land in West Africa: A Synthesis of Findings from Research on Derived Rights to Land.* London: International Institute for Environment and Development.

Leach, Melissa, and Robin Mearns, eds. 1996. *The Lie of the Land: Challenging Received Wisdom on the African Environment.* Oxford, UK: James Currey.

LEGS (Livestock Emergency Guidelines and Standards) Steering Group. 2008. "Livestock Emergency Guidelines and Standards." http://www.livestock-emergency.net/.

Lesorogol, Carolyn K. 1998. "Life on the Margins: Perspectives on Pastoralism and Development in Kenya." Unpublished report to the United States Agency for International Development, Washington, DC.

Lewis, John Van Dusen. 1979. "Descendants and Crops: Two Poles of Production in a Malian Peasant Village." Unpublished PhD diss., Yale University, New Haven, CT.

Lister, Sarah. 2004. "The Processes and Dynamics of Pastoral Representation in Ethiopia." Working Paper 22, Institute of Development Studies, University of Sussex, Brighton, U.K.

MA (Millennium Ecosystem Assessment). 2005. *Ecosystems and Human Well-Being: Desertification Synthesis.* Washington, DC: World Resources Institute.

Markakis, John. 2004. *Pastoralists on the Margin.* London: Minority Rights Group.

Mendes, Lloyd. 1988. "Private and Communal Land Tenure in Morocco's Western High Atlas Mountains: Complements, Not Ideological Opposites." Pastoral Development Network Paper 26a, Overseas Development Institute, London.

Mitchell, Donald. 2008. "A Note on Rising Food Prices." Policy Research Working Paper 4682, World Bank, Washington, DC. http://www-wds.worldbank .org/servlet/WDSContentServer/WDSP/IB/2008/07/28/000020439_200807281 03002/Rendered/PDF/WP4682.pdf.

Morgan, Jack A., Daniel G. Milchunas, Daniel R. LeCain, Mark West, and Arvin R. Mosier. 2007. "Carbon Dioxide Enrichment Alters Plant Community Structure and Accelerates Shrub Growth in the Shortgrass Steppe." *Proceedings of the National Academy of Sciences* 104 (37): 14724–29.

Mortimore, Michael J., and William M. Adams. 2001. "Farmer Adaptation, Change and 'Crisis' in the Sahel." *Global Environmental Change* 11 (1): 49–57.

Morton, John. 1988. "Sakanab: Information and Greetings among the Northern Beja." *Africa* 58 (4): 2–7.

———. 2002. "Drought Management as Growth Industry and as Development Paradigm." Paper presented at the Development Studies Association Annual Conference, University of Greenwich, U.K., November 9.

———. 2006. "Pastoralist Coping Strategies and Emergency Livestock Market Intervention." In *Pastoral Livestock Marketing in Eastern Africa: Research and Policy Challenges,* ed. John McPeak and Peter D. Little. Bourton-on-Dunsmore, U.K.: Intermediate Technology Publications.

———. 2007. "The Impact of Climate Change on Smallholder and Subsistence Agriculture." *Proceedings of the National Academy of Sciences* 104: 19680–85.

Morton, John, John K. Livingstone, and Mohammed Mussa. 2007. "Legislators and Livestock: Pastoralist Parliamentary Groups in Ethiopia, Kenya and Uganda." *IIED Gatekeeper* 131. London: International Institute for Environment and Development.

Oram, Peter. 1998. "The Influence of Government Policies on Livestock Production and the Environment in West Asia and North Africa (WANA)." In *Livestock and the Environment: International Conference,* ed. Arend J. Nell. Wageningen, The Netherlands: International Agricultural Centre.

Reij, Chris. 2008. "Promoting the Re-Greening of the Sahel." Unpublished manuscript, Free University of Amsterdam, Netherlands.

Safriel, Uriel, and Zafar Adeel. 2005. "Drylands Systems." In *Ecosystems and Human Well-Being, Volume 1—Current State and Trends, Findings of the Conditions and Trends Working Group,* ed. Rashid M. Hassan, Robert Scholes, and Neville Ash, 623–62. Washington, DC: Island Press.

Scoones, Ian, and William Wolmer. 2005. *Pathways of Change in Africa: Crops, Livestock and Livelihoods in Mali, Ethiopia and Zimbabwe.* Oxford, U.K.: James Currey.

SOS Sahel. 2008. "Biofuels, Jatropha and Pastoralism." Background paper prepared for the High-Level Workshop on Biofuels and Pastoralists, Oxford, U.K., October 22.

Swearingen, Will D., and Abdellatif Bencherifa. 2000. "An Assessment of the Drought Hazard in Morocco." In *Drought: A Global Assessment Volume I,* ed. Donald A. Wilhite, 279–86. London: Routledge.

Swift, Jeremy. 1996. "Desertification: Narratives, Winners and Losers." In *The Lie of the Land: Challenging Received Wisdom on the African Environment,* ed. Melissa Leach and Robin Mearns, 73–90. Oxford, U.K.: James Currey.

———. 2001. "District-Level Drought Contingency Planning in Arid Districts of Kenya." In *Pastoralism, Drought and Planning: Lessons from Northern Kenya and Elsewhere,* ed. John Morton, 40–84. Chatham, U.K.: National Resources Institute.

Thornton, Philip, Peter G. Jones, Tom Owiyo, Russ Kruska, Mario Herrero, Patti Kristjanson, An Notenbaert, Nigat Bekele, and Abisalom Omolo, with contributions from Victor Orindi, Brian Otiende, Andrew Ochieng, Suruchi Bhadwal, Kadambari Anantram, Sreeja Nair, Vivek Kumar, and Ulka Kulkar. 2006. "Mapping Climate Vulnerability and Poverty in Africa." Report to the Department for International Development, International Livestock Research Institute, Nairobi, Kenya.

Toulmin, Camilla. 1983. "Economic Behaviour among Livestock-Keeping Peoples: A Review of the Literature on the Economics of Pastoral Production in the Semi-Arid Zones of Africa." Development Studies Occasional Paper 25, University of East Anglia, Norwich, U.K.

———. 1992. *Cattle, Women and Wells: Managing Household Survival in the Sahel.* Oxford, U.K.: Clarendon Press.

———. 2009. *Climate Change in Africa.* London: Zed Books.

Toulmin, Camilla, Rebecca Leonard, Ngolo Coulibaly, Grace Carswell, and Data Dea. 2000. "Diversification of Livelihoods: Evidence from Mali and Ethiopia." Research Report 47, Institute of Development Studies, University of Sussex, Brighton, U.K.

Trenberth, Keven E., and Philip D. Jones. 2007. "Observations: Surface and Atmospheric Climate Change." In *Climate Change 2007: The Physical Science Basis. Contribution of Working Group I to the Fourth Assessment Report of the Intergovernmental Panel on Climate Change,* ed. Susan Solomon, Dahe Qin, Martin Manning, Melinda Marquis, Kristen Averyt, Melinda M. B. Tignor, Henry LeRoy Miller Jr., and Zhenlin Chen, 235–336. Cambridge, U.K.: Cambridge University Press.

UNDP (United Nations Development Programme). 2001. "Summary Paper by India." Paper prepared for the Regional Seminar on Drought Mitigation, Tehran, Islamic Republic of Iran, August 28–29. http://www.reliefweb.int/rw/RWFiles2001.nsf/FilesByRWDocUNIDFileName/OCHA-64BQYF-undp_irn2_29aug.pdf/$File/undp_irn2_29aug.pdf.

van Duivenbooden, N., S. Abdoussallam, and A. Ben Mohamed. 2002. "Impact of Climate Change on Agricultural Production in the Sahel. Part 2: Case Study for Groundnut and Cowpea in Niger." *Climatic Change* 54 (3): 349–68.

Wiggins, Steve, and Junior Davis. 2006. "Economic Institutions." Briefing Paper 3, Economic Institutions, Research Programme Consortium on Improving Institutions for Pro-Poor Growth, Institute for Development Policy and Management, University of Manchester, U.K.

Wiggins, Steve, Enrica Fioretti, Jodie Keane, Yasmeen Khwaja, Scott McDonald, Stephanie Levy, and C. S. Srinivasan. 2008. "Review of the Indirect Effects of Biofuels: Economic Benefits and Food Insecurity." Report to the Renewable Fuels Agency. Overseas Development Institute, London.

Wilbanks, Tom, Patricia Romero Lankao, Manzhu Bao, Frans Berkhout, Sandy Cairncross, Jean-Paul Ceron, Manmohan Kapshe, Robert Muir-Wood, and Ricardo Zapata-Marti. 2007. "Industry, Settlement and Society." In *Climate Change 2007: Impacts, Adaptation and Vulnerability. Contribution of Working Group II to the Fourth Assessment Report of the Intergovernmental Panel on Climate Change,* ed. Martin L. Parry, Osvaldo F. Canziani, Jean Palutikof, Paul van der Linden, and Clair Hanson, 357–90. Cambridge, U.K.: Cambridge University Press.

CHAPTER 9

Toward Pro-Poor Adaptation to Climate Change in the Urban Centers of Low- and Middle-Income Countries

Caroline Moser and David Satterthwaite

To date, the need to begin addressing climate change risks in the urban areas of low- and middle-income countries is not appreciated fully by most governments and the majority of development and disaster specialists (Satterthwaite et al. 2007). Low- and middle-income countries not only account for nearly three quarters of the world's urban population (United Nations 2008), but also have most of the urban populations at greatest risk from the increased intensity and/or frequency of storms, flooding, landslides, and heat waves that climate change is bringing or will bring (United Nations 2008; Wilbanks et al. 2007). Since 1950, there has been a sevenfold increase in their urban populations. This increase also has brought an increased concentration of people and economic activities in low-lying coastal zones at risk from sea-level rise and extreme weather events (McGranahan, Balk, and Anderson 2007). Globally, most deaths from disasters related to extreme weather occur in these countries, with a large and growing proportion of such deaths in urban areas (UN-Habitat 2007).

Low- and middle-income countries also have much less adaptive capacity than do high-income countries because of backlogs in protective infrastructure and services as well as local government limitations. This problem is compounded by the unwillingness of many city and municipal governments to work with the residents of informal settlements, even though these settlements often house a third or more of the population and include those people who are most at risk from climate change. The limits in local governments' adaptive capacities have led to recognition of the importance of adaptive capacity for low-income individuals, households, and communities within these settlements. Equally, it has led to recognition of the need to support initiatives that build household and community resilience and that adapt assets and capabilities so that they are able to cope with climate change. Thus, addressing the social dimensions of climate change adaptation requires that we consider the roles not only of different levels of government but also of individuals, households, and civil society organizations. This chapter seeks to address these issues by outlining a framework of pro-poor asset adaptation for climate change. This framework provides a conceptual approach for identifying the asset vulnerability to climate change of low-income individuals, households, and communities; and it considers how assets can support adaptation. Such an approach recognizes that strengthening the asset base of low-income households and communities also can contribute to building more competent, accountable local governments. A substantial part of adaptive capacity relates to the ability of local communities to make demands on local governments and, wherever possible, to work in partnership with them.

Background: The Urgency of Climate Change in Urban Contexts

Table 9.1 summarizes some of the main manifestations and likely impacts of climate change. These include increased frequency and/or intensity of extreme weather events, including heat waves; heavy precipitation; and intense, tropical cyclones. Extreme weather events long have been among the most common causes of disasters, independent of climate change. There has been a clear upward trend in the frequency of disasters from 1950 to 2005, and especially from 1980 (UN-Habitat 2007); and most of this upward trend is the result of events related to extreme weather (Hoeppe and Gurenko 2007). The growing number of extreme weather-related disasters is consistent with predictions from the Intergovernmental Panel on Climate Change (IPCC) of what climate change will bring. It is not "proof of climate change" (which is difficult to ascertain), but proof of the vulnerability of cities and smaller settlements to extreme weather events whose frequency and intensity climate change is likely to increase.

Even though this chapter focuses on urban areas, agriculture is included in table 9.1, not only because of its influence on the price and availability of food, fuel, and many industrial inputs for urban areas; but also because

Table 9.1. Likely Impacts of Climate Change

Change	Impact on Natural Systems, Agriculture, Water	Impact on Urban Areas	Impact on Health and Household Coping	Implications for Children
Increased frequency of warm spells and heat waves over most land areas	Reduced crop yields in warmer regions, increased wildfire risk, wider range for disease vectors	Heat islands with higher temperatures (up to 7°C higher); often large concentrations of vulnerable people; worsened air pollution	Increased risk of heat-related mortality and morbidity, more vector-borne disease, impacts for those doing strenuous labor, increased respiratory disease where air pollution worsens, food shortages	Greatest vulnerability to heat stress for young children; high vulnerability to respiratory diseases, vector-borne diseases; highest vulnerability to malnutrition, with long-term implications
Increased frequency of heavy precipitation events over most areas	Damage to crops, soil erosion, waterlogging, water quality problems	Floods and landslide risks increase; disruption to livelihoods and city economies; damage to homes, possessions, businesses, transport, and infrastructure; loss of income and assets; often large displacements of population, with risks to social networks and assets	Deaths, injuries, and increased food-borne, water-borne, and water-washed diseases; more malaria from standing water; decreased mobility with implications for livelihoods; dislocations; food shortages; risks to mental health, especially associated with displacement	Higher risk of death and injury than adults; more vulnerable to water-borne and water-washed illness and to malaria; risk of acute malnutrition; reduced options for play and social interaction; likelihood of being removed from school/put to work as income is lost; higher risk of neglect, abuse, and maltreatment associated with household stress and/or displacement; long-term risks for development and future prospects
Increases in intense tropical cyclone activity	Damage to crops, trees, and coral reefs; disruption to water supplies			

(continued)

Table 9.1. Likely Impacts of Climate Change (*continued*)

Change	Impact on Natural Systems, Agriculture, Water	Impact on Urban Areas	Impact on Health and Household Coping	Implications for Children
Increased area affected by drought	Land degradation, lower crop yields, livestock deaths, wildfire risks, and water stress increase	Water shortages, distress migration into urban centers, hydroelectric constraints, lower rural demand for goods and services, higher food prices	Increased shortages of food and water, increased malnutrition and food- and water-borne diseases, elevated risk of mental health problems, respiratory problems from wildfires	Young children at highest health risk from inadequate water supplies; at highest risk of malnutrition, with long-term implications for overall development; risk of exploitation and early entry into work
Increased incidence of extremely high sea level	Salinization of water sources	Loss of property and enterprises, damage to tourism and damage to buildings from rising water table	Coastal flooding, increasing risk of death and injuries; loss of livelihoods; health problems from salinated water	Highest rates of death for children; highest health risks from salinization of water supplies, with long-term developmental implications

Sources: IPCC (2007), table SPM.1; and Bartlett (2008).

of the importance of rural-based (producer and consumer) demand for goods and services for many urban economies. In addition, there are important rural-urban links for adaptation, such as the protection of key natural defenses within and around urban centers and watershed management linked to flood control and protection of the water supply.

The fourth assessment of the IPCC notes that "climate change can threaten lives, property, environmental quality and future prosperity by increasing the risk of storms, flooding, landslides, heat waves and drought and by overloading water, drainage and energy supply systems" (Wilbanks et al. 2007, p. 382). Many urban centers will also be affected by less dramatic stresses, including reductions in freshwater availability and stresses on local agricultural production.

It is difficult to generalize about likely risks of climate change because the scale and nature of these risks vary greatly between urban centers, and between different population groups or locations within urban centers. Nevertheless, these centers can be grouped according to certain shared physical characteristics that relate to climate change risk:

- already facing serious impacts from heavy rainstorms and cyclones (including hurricanes and typhoons) or heat waves
- coastal location, affected by sea-level rise
- location by a river that may flood more frequently
- dependent on freshwater sources whose supply may diminish or whose quality may be compromised.

The extent to which extreme weather events and other likely climate change impacts pose problems relates not only to settlement location, but also to the quality and level of infrastructure and service provision. A high proportion of deaths, serious injuries, and loss of property from storms, floods, and landslides is the result of deficiencies in such provision. This also means that there are large variations in the relative importance of climate change risks, compared with other pressing environmental hazards. Where a large proportion of an urban center's population lacks infrastructure—such as piped water, sanitation, and drainage—it is difficult to claim the problem is primarily one of climate change. In addition, the very large variations in the numbers of people killed or injured by extreme weather events is much influenced by the quality and extent of protective infrastructure. Wealthy cities and nations can afford levels of investment in protective infrastructure that are far beyond those possible in low- and middle-income nations.

A Conceptual Framework: From Asset Vulnerability to Asset Adaptation

A social development perspective on urban climate change adaptation focuses on both the impacts and the risks these bring for the low-income, excluded, and marginalized populations. Recognition that low-income populations are particularly vulnerable to climate change, in terms of individual lives and in relation to their household and community assets, makes it useful to build on earlier conceptual and operational frameworks on poverty, vulnerability, and assets (Moser 1998, 2007, 2008); and to modify these in addressing the particular problems associated with climate change.

Asset Vulnerability

Analysis of the risks arising from climate change to low-income urban households and communities is grounded in the concept of vulnerability. This concept draws on an important body of development literature that recognizes poverty as more than income or consumption poverty and that captures the multidimensional aspects of changing socioeconomic well-being. In an urban study, Moser (1998) defines vulnerability as insecurity in the well-being of individuals, households, and communities, including sensitivity to change. Vulnerability can be understood in terms of a lack of resilience to a variety of changes that threaten welfare. These changes can be environmental, economic, social, and political. They can take the form of sudden shocks, long-term trends, or seasonal cycles. Such changes usually bring increasing risk and uncertainty. Although the concept of vulnerability has focused mainly on its social and economic components, in applying it to climate change, vulnerability to physical hazards, certain diseases, and heat stress become more important.

Also of operational relevance to adaptation is the distinction between vulnerability and capacity/capability, which is linked to resilience. The emergency relief literature has shown that people are not "helpless victims," but have many resources even at times of emergency; and that these resources should form the basis for responses (ACHR 2005; Longhurst 1994). There also is a growing recognition of the resources that grassroots organizations can bring to urban adaptation (Huq and Reid 2007; Satterthwaite et al. 2007).

The fact that vulnerability can be applied to a range of hazards, stresses, and shocks offers a particular advantage to the analysis of climate change–related risks. Urban poor populations generally have to live with multiple risks and have to manage the costs and benefits of overlapping hazards from a range of environmental sources, while they face economic, political, and social constraints. Climate change also brings a futures dimension to understanding vulnerability. It highlights the uncertainty of future risk and, with this, an uncertainty concerning the bundle of assets that will enable adaptation and increased resilience. An asset-based vulnerability approach that incorporates social, economic, political, physical, human, and environmental resources allows for flexibility in analysis and in planning interventions that is harder to maintain within a hazard-specific approach. It also highlights how many assets serve to reduce vulnerability to a range of hazards.

Generally, the more assets people have, the less vulnerable they are; the greater the erosion of people's assets, the greater their insecurity. An asset is identified as a "stock of financial, human, natural or social resources that can be acquired, developed, improved and transferred across generations. It generates flows of consumption, as well as additional stock" (Ford Foundation 2004, p. 9). In the current poverty-related development debates, the concept of assets or capital endowments includes both tangible and intangible assets—with the assets of the poor commonly identified as natural, physical, social, financial, and human capital. In postdisaster impact assessments, assets are shown both to be a significant factor in self-recovery and to be influenced by the response and reconstruction process. (See the "Community Responses to Climate Change" section below.)

Asset-Based Adaptation

Assets "are not simply resources that people use to build livelihoods: they give them the capability to be and act" (Bebbington 1999, p. 2029). As such, assets are identified as the basis of agents' power to act to reproduce, challenge, or change the rules that govern the control, use, and transformation of resources (Sen 1997). Moser (2007) distinguishes between an asset-index conceptual framework, which she defines as a diagnostic tool for understanding asset dynamics and mobility; and an asset-accumulation policy, which she defines as an operational approach for designing and implementing sustainable asset-accumulation interventions.

To get beyond vulnerability and focus on strategies and solutions, this chapter introduces an asset-based framework that identifies the role of assets in increasing the adaptive capacity of low-income households and communities to climate change. Asset-based frameworks include a concern for long-term accumulation strategies (see Carter 2007 and Moser 2007).

Clearly the asset portfolios of individuals, households, and communities are a key determinant of their adaptive capacity, both to reduce risk and to cope with and adapt to increased risk levels.

An asset-based adaptation strategy in the context of climate change includes three basic principles:

1. The process by which the assets held by individuals and households are protected or adapted does not take place in a vacuum. External factors such as government policy, political institutions, and non-governmental organizations all have important roles. (Again, see the "Community Responses to Climate Change" section below.) However, institutions can also include the laws, norms, and regulatory frameworks that block, enable, or positively facilitate asset adaptation.

2. The formal and informal context within which actors operate can provide an enabling environment for protecting or adapting assets. Entry points for strengthening strategies for asset adaptation are contextually specific but may be institutional- or opportunity-related in focus. Just as the adaptation of one asset often fosters the adaptation of other assets, the insecurity and erosion of one asset can undermine the stability of other assets.

3. Household asset portfolios change over time, sometimes rapidly—for example, following events such as marriage or the death or incapacity of an income earner. Thus, households quickly can move into insecurity/vulnerability through internal changes linked to life cycle as well as in response to external economic, political, and institutional change.

Thus, key to the development of an asset-based adaptation framework are the identification and analysis of the connection between vulnerability and the erosion of assets.

Urban Poverty, Asset Vulnerability, and Climate Change

To address vulnerability, it is important to understand how vulnerability to different aspects of climate change varies across different groups (for instance, not only by type of hazard, but also by age, gender, health status, and income); and how asset portfolios can influence this variability. Furthermore, it is important to consider not only risk and vulnerability to events that threaten life, health, and livelihood; we also must consider risks and vulnerabilities in the aftermath of such events.

Types of Vulnerability and Groups Particularly Affected

Hazards created or magnified by climate change combine with vulnerabilities to affect the urban poor's human capital (health), physical capital (housing and capital goods), and capacity to generate financial and productive assets. Some impacts are direct, such as more frequent and more intense floods. Those effects that are less direct include reduced freshwater supplies. Finally, others that are indirect for urban populations include constraints on agriculture that, in turn, constrain food supplies and reduce rural producers' and consumers' demands for goods and services from urban enterprises.

There is considerable variation in levels of vulnerability within low-income populations, in terms of both the hazards to which they are exposed and their capacity to cope and adapt. Settlements differ in the quality of physical capital (including housing) and infrastructure (much of which should reduce risks) and in the risks to which they are exposed from flooding, landslides, and heat waves. In addition, a local population's interest in risk-reduction through building improvements will vary depending on home ownership status, because tenants or seasonal migrants have fewer incentives to invest.

There also are differences between women's and men's exposures to hazards and their capacities to avoid, cope with, or adapt to them. Age, too, is important, with children facing greater risks and having reduced coping capacities for some impacts; very young children and older groups face particular risks from some effects. Individual health status also is crucial, regardless of age and gender (Bartlett 2008).

Areas of Intervention

To look in more detail at the connections between vulnerability and erosion of assets, this chapter considers one of the most important aspects of climate change—the likely increase in the intensity and/or frequency of floods and storms. It addresses vulnerability within four different areas of intervention: protection, predisaster damage limitation, immediate postdisaster response, and rebuilding. Before discussing these interventions, it is important to note that many components of poverty reduction strategies also build resilience against a range of hazards, and thus complement actions targeted at particular groups' exposures to specific hazards. For instance, better-quality housing, infrastructure, and services greatly reduce a range of hazards—including exposure to many physical dangers and disease-causing agents (pathogens)—while an effective health care system should reduce the impact of most illnesses and injuries.

Protection. Among the groups most at risk are lower-income groups living in environmentally hazardous areas, lacking protective infrastructure—for instance, large concentrations of illegal settlements on hills prone to landslides, in deep ravines, or on land likely to flood (Hardoy, Mitlin, and Satterthwaite 2001). Most major cities were founded on "safe" sites, but have grown to sizes never imagined by their founders. Increased exposure to extreme weather hazards is related partly to expansion onto hazardous sites. Risks faced on such sites have often been exacerbated by damage to natural systems, including the loss of mangrove stands or hillside vegetation as well as deforestation. But areas constantly exposed to flooding still attract low-income groups because of cheaper land on which they can build or cheaper rental accommodations.

Extreme weather impacts frequently relate more to the lack of protective infrastructure and services than to the hazards inherent in urban sites (Revi 2008). For instance, it is generally cities with the largest inadequacies in protective infrastructure that have experienced the highest number of flood-related deaths and injuries over the past 25 years. The lack of protective infrastructure is partly linked to the constrained investment capacity of city and municipal governments. But it is often associated with the problematic relationships between local governments and urban poor groups living in high-risk, informal settlements. High proportions of these cities' populations occupy land illegally, build structures that contravene building regulations, and work in the informal economy outside official rules and regulations. Infrastructure and service-provision agencies may not work in such informal settlements because of the "anti-poor" attitudes of government officials and politicians. Within such settlements, most loss or damage to buildings in extreme weather is the result of inadequate infrastructure (for instance, for storm drainage) and poor building quality.

Predisaster Damage Limitation. Generally, high-income groups and formal businesses with good-quality buildings and safe, protected sites do not need to take emergency preparedness measures in response to forecasts for storms and high tides. For groups living in less resilient buildings and more dangerous sites, risks to health and assets can be reduced by appropriate actions in response to warnings. However, to be effective, reliable information must reach in advance those most at risk, be considered credible, and contain supportive measures that enable them to take risk-reducing actions. These measures include the identification of known safer locations and, where needed, provision of transport to assist people's moving to them.

Climate change is likely to make the timing and intensity of heavy rainfall less predictable, which, in turn, makes long-established coping mechanisms less effective. Discussions with residents of low-income communities in various African cities suggests that flooding has become more frequent and less predictable (Douglas et al. 2008). Effective community-based predisaster measures to limit damage also require levels of trust and cohesion—community social capital—that are often not present. These measures depend on a complex set of factors, including length of time in the settlement, pattern of occupation (including tenure), and state infrastructure-delivery mechanisms (see Moser and Felton 2007).

There also are differences in knowledge and the capacity to act to limit risk, based on age, gender, and health status. These include differentials as simple as the capacity to run or to swim, with speed variations relating to different groups; infants, younger children, adults caring for them, the disabled, and older people all move more slowly when responding to impending risks. It is common for mortality among children to be higher than among adults (Bartlett 2008). In societies where women are constrained by social norms from leaving the home, they may move less rapidly to avoid floodwater, as may women who are taking responsibility for young children.

Immediate Postdisaster Response. This intervention concerns groups less able to cope with impacts. Disasters often separate communities, inhibiting responses by established community organizations. Particular groups, differentiated by age, gender, health status, and other forms of exclusion such as ethnicity or religion, face particular difficulties in coping with the immediate effects. Infants, young children, and members of older age groups are at higher risk from postdisaster disruptions such as scarcities of potable water and food. When income is lost, children from poor households may be removed from school to help provide for the household. Disaster events often endanger the personal safety of girls and women, with higher risk of gender-based violence, abuse, and maltreatment associated with displacement and/or household stress (Bartlett 2008). Although little is known about the psychological impact of urban disasters, it is clear that different forms of trauma unfold over time, from acute shock lasting a few days, to longer-term impacts such as recurrent stress-related illnesses and reduced quality of life (Bartlett 2008).

Rebuilding. Poorer groups generally receive less support from the state for rebuilding and very rarely have insurance protection that helps fund the rebuilding of homes and other assets. Postdisaster reconstruction processes rarely allow the poorest groups and those most affected to take central roles in determining locations and forms of reconstruction. In many instances, poor groups do not get back the land from which they were displaced because the land is declared a no-build zone by governments or is acquired by commercial developers (ACHR 2005).

Women generally assume most childrearing and domestic responsibilities such as obtaining food, fuel, and water—all of which can become more burdensome following a disaster. At the same time, they "struggle in the fast-closing post-disaster 'window of opportunity' for personal security, land rights, secure housing, employment, job training, decision-making power, mobility, autonomy, and a voice in the reconstruction process" (Enarson and Meyreles 2004, p. 69). Equally problematic is the failure to recognize women's individual and collective recovery and reconstruction capacities as community leaders, neighborhood networkers, producers, gardeners, rainwater harvesters, and monitors of flood-prone rivers (Enarson and Meyreles 2004; Enarson et al. 2003). Children also generally are affected in more extreme ways because of their higher physiological and psychosocial vulnerability to a range of associated stresses, as well as the long-term developmental implications of these vulnerabilities. Almost all of the disproportionate implications for children are exacerbated by poverty and by the difficult choices that must be made by low-income households as they adapt to more challenging postdisaster conditions (Bartlett 2008).

Current Governmental Operational Frameworks for Action

At first sight, climate change adaptation frameworks seem primarily to be a municipal government responsibility, with limited roles for households and community organizations. City and municipal governments are in charge of planning, implementing, and managing most measures that can diminish climate change risks, including those that address particular groups' high vulnerability to known hazards. Their responsibilities also include factoring climate change risks into new development plans and investment programs, and adapting infrastructure standards and building codes.

Urban populations in high-income nations take for granted that a web of institutions, infrastructure, services, and regulations protects them from extreme weather and floods, and keeps adapting to continue protecting them. Many measures to protect against extreme weather meet everyday needs—for instance, health care services are integrated with emergency services and sewer and drainage systems serve daily requirements and cope with heavy rainfall. The police, armed services, health services, and fire services provide early warning and ensure rapid emergency responses. The costs of these efforts are recovered by administering service charges or taxes; and, for most people, these costs represent a small proportion of their income. Consequently, extreme weather events rarely cause large loss of life or serious injury in high-income nations. Although such events occasionally cause serious property damage, the economic cost is reduced for most property owners by property and possessions insurance.

This adaptive capacity is supported further by most buildings conforming to building and health and safety regulations, and being served at all times by piped water, sewers, all-weather roads, electricity, and drains. The institutions responsible for such services are expected to make them resilient to extreme weather. Although private companies or nonprofit institutions may provide some of the key services, the framework for provision and quality control is supplied by local government or local offices of provincial or national government. In addition, it is assumed that city planning and land-use regulation will be adjusted to any new or heightened risk that climate change may bring, encouraged and supported by changes in private sector investments (over time, shifting from high-risk areas) and changes in insurance premiums and coverage. At least for the next few decades, as the IPCC's fourth assessment stresses, this "adaptive capacity" can deal with most likely impacts from climate change in the majority of urban centers in high-income countries.

For the most part, households and community organizations in high-income countries engage very little with the institutions that ensure their protection, other than through complaint channels such as local politicians, lawyers, ombudsmen, and consumer groups and watchdogs. Some groups may be ill served or excluded, but most urban inhabitants are well served and protected.

Very few urban centers in low- and middle-income nations have a comparable web of institutions, infrastructure, services, and regulations. In urban centers where much of the population lives in informal settlements, not only is the public provision of infrastructure and services inadequate, but there are also few mechanisms by which low-income citizens can hold their local governments accountable. At the same time, many such local governments are anti-poor, regarding informal-settlement populations as "the problem" rather than as key parts of the urban economy. There are urban centers where deficiencies in infrastructure and services affect a smaller proportion of the population; and in such centers, this often reflects local governments that are more accountable to the citizens in their jurisdictions, with national government structures that have supported decentralization. Progress in such urban centers often is associated with stronger local democracies (Cabannes 2004; Campbell 2003; Velásquez 1998). Some local governments have developed successful partnerships with low-income groups and their community organizations that demonstrate cheaper, more effective ways to meet their responsibilities for infrastructure and services (Mitlin 2008; Hasan 2006; Boonyabancha 2005; d'Cruz and Satterthwaite 2005).

Community Responses to Climate Change: A Pro-Poor, Asset-Based Adaptation Framework for Storms and Floods

If most city or municipal governments have proved unable or unwilling to provide the infrastructure, services, institutions, and regulations to reduce risks from extreme weather events for much of their populations, they are unlikely to develop the capacity necessary to adapt to climate change. Adaptation frameworks need to be developed to support household- and community-based responses. This development might be considered as support for adaptation that is independent of government, but support for household and community adaptation also should be supporting citizens' capacity to negotiate and work with government wherever possible and, if needed, to contest government. Table 9.2 provides examples of asset-based actions at different levels to build resilience to extreme weather.

Obviously, the greater the success in protection, the less the need for intervention in the second, third, and fourth aspects discussed above. Similarly, good predisaster damage limitation greatly can reduce the impacts and the scale of the required postdisaster response and rebuilding. This is a critical point, with implications for human well-being (including lives saved

Table 9.2. Examples of Asset-Based Actions at Different Levels to Build Resilience to Extreme Weather

Areas of intervention	Asset-Based Actions		
	Household and Neighborhood	Municipal/City	Regional or National
Protection	• Take household and community-based actions to improve housing and infrastructure • Conduct community-based negotiation for safer sites in locations that serve low-income households • Take community-based measures to build disaster-proof assets (such as savings) or protect assets (for example, insurance)	• Work with low-income communities to support slum and squatter upgrading, informed by hazard mapping and vulnerability analysis • Support increased supply and reduced costs of safe sites for housing	• Develop government frameworks to support household, neighborhood, and municipal action • Make risk-reducing investments and take actions that are needed beyond urban boundaries
Predisaster damage limitation	• Develop community-based disaster preparedness and response plans, including early-warning systems that reach everyone, measures to protect houses, safe evacuation sites identified if needed, and provisions to help those less able to move quickly	• Install early-warning systems that reach and serve groups most at risk • Prepare safe sites with services, and organize transportation to safe sites • Protect evacuated areas from looting	• Establish national weather-monitoring systems capable of providing early warning • Support community and municipal actions
Immediate postdisaster response	• Support immediate household and community responses to reduce risks in affected areas • Support the recovery of assets • Develop and implement responses	• Encourage and support active engagement of survivors in decisions and responses • Draw on resources, skills, and social capital of local communities • Rapidly restore infrastructure and services	• Fund and provide institutional support for community and municipal responses
Rebuilding	• Provide support for households and community organizations to get back to their homes and communities • Plan for rebuilding with greater resilience • Provide support for recovering the household and local economy	• Ensure reconstruction process supports household and community actions, including addressing priorities of women, children, and youth • Build or rebuild infrastructure to more resilient standards	• Fund and provide institutional support for household, community, and municipal action • Address deficiencies in regional infrastructure

Source: Authors' compilation.

and injuries and asset losses avoided) and for economic costs. Promoting the protection intervention to politicians and civil servants is hampered by the difficulty of enumerating the lives it will save and the injuries it will avoid. However, some idea of these numbers can be seen in the mortality differences among various cities hit by cyclones of comparable strength. There has been more analysis of the economic savings from disaster prevention than of the costs of reconstruction, and it highlights the very large cost advantages of protection and disaster prevention (ODI 2004).

Asset-Based Adaptation for Protection

Protecting physical assets for low-income groups focuses on safer residential sites, better housing, and protective infrastructure. In the majority of instances, the most effective adaptation is establishing the infrastructure and institutions that prevent storms or floods from becoming disasters. For most urban centers in low- and middle-income nations, this intervention also is the most difficult to implement, primarily because of the lack of funding, the government's incapacity, and the large deficits in infrastructure provision. In many nations, the difficulty in implementing the intervention also relates to how higher levels of government have retained the power, resources, and fundraising capacities that urban governments need.

It is important to recognize that most low-income urban groups already have a range of measures by which they adapt to risk and to changing circumstances, whether these are economic opportunities or shocks, political circumstances, or housing risks. However, their survival needs and economic priorities often conflict with risk reduction (see, for instance, Stephens, Patnaik, and Lewin 1996). A study of disaster-prone "slum" communities in El Salvador showed how the individualistic nature of households' investments, the lack of representative community organizations, and the lack of appropriate support from governments and local and international nongovernmental organisations inhibited needed settlementwide risk reduction measures (Wamsler 2007).

There is also the issue of what community organizations cannot address. Much of the needed protection in cities is for large-scale, expensive infrastructure that is part of citywide systems—for instance, storm and surface drains (and measures to keep them free of silt and solid waste). The scale and range of what community-based organizations can achieve in developing protective infrastructure are much increased where they can work in partnership with government agencies (see, for instance, Hasan 2006).

The relocation of existing houses and settlements away from areas that cannot be protected from floods and storms, coupled with land-use management strategies to prevent new settlements in such areas, are important components of an asset-based strategy. However, homeowners and renters will often resist relocation because it can result in a decline in financial capital and social networks, as well as loss of the physical asset itself: the housing. For poor urban households, housing is the first and most important asset they seek to acquire (see Moser and Felton 2007). Climate change will often reduce the availability of safe residential sites because it increases the sites at risk from mudslides, wind damage, flooding, and (for coastal cities) rising sea levels. In many nations, however, there are examples of low-income households obtaining safer, legal sites for their housing as a result of the active engagement of these households in organizations or federations of "slum" or shack dwellers (Manda 2007; Sisulu 2006; Boonyabancha 2005; Burra, Patel, and Kerr 2003). Although those examples were not driven by climate change adaptation, they demonstrate how relocation agreements were reached between very-low-income households and governments in a relocation methodology that involved poorer groups, thus avoiding many of the disadvantages and the impoverishment that often accompanies resettlement.

Asset-Based Adaptation for Predisaster Damage Limitation

Most urban centers in low- and middle-income countries at high risk from extreme weather lack the capacity to invest in measures that provide complete protection. In such circumstances, well-conceived interventions made in the period just prior to the extreme event can greatly reduce loss of life, serious injury, and loss of possessions; and can have the potential to moderate damage to homes.

One of the foundations of predisaster damage limitation is an early-warning system that not only identifies the risk, but also communicates the information to all neighborhoods at risk and supports credible, realistic responses. This is not something that low-income communities can provide for themselves; rather, it depends on government institutions. Many low-income countries do not have an adequate weather-monitoring system, although the importance of such a system now is recognized more widely. A warning system alone does not generate the required responses, however. That is especially true if low-income communities distrust local governments; are frightened that, if they move, they will not be allowed back to their settlement (because it is illegal); or know that there is likely to be looting (see Hardoy and Pandiella 2009).

There are examples of government providing early warning and support for immediate predisaster action that enabled individuals, households, and communities to take appropriate action to limit damage (United Nations 2009). One key underpinning of these examples is redefining the causes of disasters, with extreme weather-related disasters now identified as failures of development rather than simply as natural events, with associated development policy shifts in avoidance and impact-reduction measures (Lavell 1999). Even though this redefinition was not actually driven by climate change, it has relevance for adaptation.

Asset-Based Adaptation for Immediate Postdisaster Response

After any disaster, two separate intervention points are the immediate response and the longer-term follow-up. The two are separated largely because responsibility for them generally is divided among different institutions, within both governments and international agencies.

One of the main influences on low-income groups' capacity to address their postdisaster needs is the effectiveness of predisaster efforts to protect their assets. Savings and savings groups can help prevent postdisaster dependency and provide a basis for reenergizing the local economy. Support for such savings groups can be an important component of community-led postdisaster response.

An awareness of the assets and capabilities of women, men, youth, and children affected by a disaster, and of their importance in immediate postdisaster response, brings changing approaches. Many of the problems experienced after disasters are related to how emergency and transitional assistance is delivered, with people frequently feeling that they have little or no control of their lives. Not only do survivors generally have no role in decisions that affect them, but also they often do not even know what decisions have been made. The resources, skills, and social capital within local communities often are overlooked in the rush to assess risks and needs.

Approaches that encourage active engagement, community control, and rebuilding social capital in the aftermath of disaster have very significant implications for children. The benefits of community-level supportive institutions for children have been well documented. Early-childhood programs, for instance, can help reduce parental stress and provide young children with a safe, structured daily routine and valuable contact with other children (Bartlett 2008; Williams, Hyder, and Nicolai 2005). Schools provide the same routine, sanctuary, and interest for older children (Nicolai and Triplehorn 2003).

Asset-Based Adaptation for Rebuilding

Where survivors participate in decision making, psychological recovery strengthens the recovery of livelihoods and well-being. Reconstruction is a period in which either entitlements can be renegotiated to improve the capacity and well-being of the poor, or poverty and inequality can be entrenched through the corresponding reconstruction of vulnerability.

Although the reconstruction process should be an opportunity to address both short- and longer-term development issues, it often just replaces old problems with new ones. There tends to be very little understanding of how reconstruction can be turned to better advantage to rebuild social as well as physical assets, and thereby contribute to poverty reduction.

The Asian tsunami of 2004 was not caused by climate change—but it showed the extreme vulnerability of coastal populations to storm surges and revealed why solid gender analysis should be included in rebuilding. After the tsunami, many women joined self-help groups to obtain micro-credit, which they used to boost their assets and increase their productive activities. This reliance on self-help groups was caused partly by the gender-blind nature of disaster relief that focused on men's lost fishing boats, not on the assets managed or controlled by women. Another tsunami lesson underscored the need to focus on rebuilding communal assets rather than individual ones. Often, individual reconstruction did not work well, and community-led development worked better. Some communities had enough power to throw out corrupt engineers or suspend them. The collective focus broke the "beneficiary" mentality, with leaders emerging to take on public roles. This response also showed how community-led reconstruction can reduce costs. Money is not wasted on unneeded infrastructure and outside professionals when the community itself has the skills to perform the necessary tasks (Moser, Sparr, and Pickett 2007).

The location of rebuilt settlements has obvious implications for livelihoods as well as for access to such amenities as schools, markets, and health facilities—and if unsuitable sites are chosen, they may remain empty (Bartlett 2008).

Recovering the household and local economy is also a cornerstone of progressive adaptation following disaster. Two core principles are required for pro-poor recovery:

1. Where possible, promote local sourcing of materials and skills to prevent monetary resources aimed at reconstruction from rapidly leaking out of the local economy.

2. Use emergency response and reconstruction interventions as a vehicle for enhancing local skills and empowerment by transferring decision-making power to survivors or sharing it with them. This moves beyond the simple employment of survivors to provide income or reduce reconstruction costs (UNDP 2004).

Local landownership and the recovery of the local economy are interdependent. Loss of rights over land and forced resettlement during reconstruction, often under the guise of "adaptation" or "risk reduction," serve to transfer land rights from the poor to the rich, while dislocating survivors from the identity of place and informal safety nets offered by social support networks.

Institutional Implications

Effective adaptation depends on the cumulative and mutually reinforcing actions and investments of a considerable range of institutions—including not only different levels of government and international agencies, but also a range of civil society organizations. It also depends on the capacity of these different institutions to work together; and for those most at risk or most vulnerable to be able to influence these institutions and hold them to account.

Implications for Urban Government

Obviously, effective adaptation strategies depend on more competent and better-resourced local governments. They also depend on local governments that are accountable and are both willing and able to work with the poor and other at-risk groups. Strong local governments that are not held locally accountable may adapt simply by evicting those people living on sites and in settlements at risk. Adaptation also depends on urban centers in which infrastructure deficiencies are much reduced, especially in informal settlements. This is a very challenging task, given the scale of these deficiencies and the lack of local capacity in many places. It is also difficult to see how the support needed for climate change adaptation can be separated from the support needed for local development, because support is needed to identify and address all environmental health risks (including everyday, small-disaster and large-disaster risks) in ways that address the risks and vulnerabilities of low-income groups and high-risk groups and that include increased resilience to the likely impacts of climate change. In effect, it is support for local development plus adaptation (Satterthwaite,

Dodman, and Bicknell 2009). Perhaps increased attention to the risks and vulnerabilities related to climate change will also serve to highlight other risks and vulnerabilities that long have been apparent but for which no action has been taken. This raises obvious questions about whether urban governments receive the needed support from national governments and international agencies.

At present, climate change models can predict likely changes at a continental or regional level, but not for particular localities; so it is not possible to predict with precision the changes that global warming will bring for each urban center. This uncertainty makes it difficult to convince local governments of any need to take action (see Roberts 2008). However, an analysis of the impacts of past extreme weather (and other disasters) can form the first step in understanding adaptation needs and in considering how to mainstream these needs into conventional planning, infrastructure investment, and other development programs (see Awuor, Orindi, and Adwera 2008 and Pelling and Wisner 2008). To such analysis should be added an information base on the current infrastructure and services provided to each building and on the details of environmental hazards—which, in turn, allows a preliminary identification of those households and areas most at risk. That information will contribute to a much more detailed and location-specific information base on risk/vulnerability, including risk-assessment maps at city and district levels. Such assessments and maps detail what is located within hazardous zones by identifying settlements, infrastructure, populations, and even gender- or age-differentiated groups most at risk. With that knowledge, choices can be made relating to investments and support programs for households and communities at high-risk sites. There also is a considerable body of experience in community-based mapping of housing, infrastructure, services, and site characteristics undertaken in informal settlements by urban poor organizations and federations (Hasan 2006; Weru 2004; Burra, Patel, and Kerr 2003). These studies allow the risk/vulnerability assessments to cover areas of the city that often include homes and neighborhoods most at risk, but for which there are little or no official data.

A pro-poor adaptation policy starts by identifying the measures to be introduced to protect those people who are identified as vulnerable. Reducing risk and increasing the resilience of physical capital that already has accumulated in cities can be done in three ways:

1. reducing hazards in occupied sites by installing protective infrastructure and complementary risk-reduction measures (which may need

modifications outside the area at risk—for instance, watershed management upstream);

2. supporting better-quality buildings—for instance, through technical support and appropriate finance systems (which also may require land tenure regularization);

3. assisting those who live in the most dangerous sites to move to safer sites, and taking measures to increase the supply and reduce the cost of land for housing on safe, serviced sites.

Implications for National Government

The potential for urban (metropolitan, city, municipal) governments to be good climate change adaptors depends heavily on the extent to which higher government levels provide the legislative, financial, and institutional basis to encourage them to do so, while not overwhelming local governments with adaptation responsibilities that cannot be fulfilled. Also important are the conditions set for urban governments applying for funding from higher levels, such as requirements for local development plans to involve all key interest groups and incorporate risk and vulnerability assessments. National funds on which innovative urban governments can draw are important. These must support locally developed responses that will vary, depending on the range and relative importance of climate change–related hazards in different urban centers. For countries where extreme weather events are already causing disasters, there is need for a national fund that supports locally developed disaster-risk reduction and rapid responses when disasters occur and helps households, civil society, and local governments in their rebuilding processes. Some obvious tasks and responsibilities for urban adaptation fall to higher government levels, such as implementing weather information systems that support local assessments and early-warning systems.

The Role of International Donors

There are three entry points for bilateral aid agencies and multilateral development banks in supporting the efforts of governments to develop adaptive capacity. The first point is examining funding flows to identify whether sufficient support is allocated to urban infrastructure and services that enhance climate resilience and disaster avoidance. The second entry point is supporting national or state/provincial-level financial and regulatory capacity to assist urban governments in developing adaptive capacity. The third point is directly supporting local adaptive capacity, working with community organizations and city and municipal governments committed to reducing the risks of climate-related hazards and able and willing to work with groups at risk. Here the focus is on linking local asset-based adaptation with good local development and environmental governance.

Increased funding flows for adaptation will not achieve much unless local governments have the capacity to use the resources appropriately and to work with the groups most at risk. For many countries, this may present more difficulties for official development assistance agencies that provide the funding. Such agencies are not set up to support the long-term local engagement necessary to ensure the development of local adaptive capacity—especially the local engagement that includes support for the asset-based adaptation frameworks so important for low-income groups and their own organizations.

An important part of building local adaptive capacity is supporting adaptation that serves low-income groups. Here there are good "slum and squatter upgrading" experiences on which to draw, where local governments worked with informal settlements' inhabitants to provide infrastructure and services and improve housing quality (see, for instance, Boonyabancha 2005). Some bilateral agencies and international foundations have developed ways to support both grassroots initiatives and the local government support for them (Mitlin and Satterthwaite 2007). Thus, those who assist adaptation must think through the financial systems and mechanisms that will enable support for a multiplicity of city or municipal innovations by households, community organizations, and local governments that reinforces and works with "good local development" and "good local governance."

New international funding sources for adaptation are being developed, beyond what development assistance agencies are already doing—especially through the United Nations Framework Convention on Climate Change. At present, the scale of funding falls far short of what is needed; and what little has been supported gives very little attention to urban areas (Ayers 2009; Satterthwaite et al. 2007). There has been some support for the least-developed nations to develop national adaptation programs of action and for community-based adaptation; but, again, little attention has been given to urban areas. In addition, such a focus gives too little attention to the key role of local government. Local adaptation programs of action are needed to underpin and drive innovations in national plans (Bicknell, Dodman, and Satterthwaite 2009). It is also important to stress that, in almost all instances, there must be "local development plus adaptation," and that much adaptation is addressing existing deficiencies in infrastructure and housing quality. In addition, competent and accountable local governments will not engage with adaptation to climate change unless it is seen as supporting and enhancing the achievement of development goals.

In addition, many international donors are concerned to see how urban adaptation also can contribute to reducing greenhouse-gas emissions (mitigation). However, measures to reduce those emissions do not necessarily serve adaptation or development. Because mitigation in high-income countries focuses so strongly on reducing use of fossil fuels, there is an assumption that the measures used to achieve this should be transferred to low- and middle-income countries, even when many countries have per capita carbon dioxide emissions that are 1/50th or even 1/100th those in high-income countries. In most urban centers in low- and middle-income countries, climate change priorities should focus on the adaptation needs of the poor, including the expansion and improvement of protective infrastructure and services and safe housing sites, not on energy efficiency. Of course, urban development in low- and middle-income nations should support mitigation—but not at the expense of adaptation or of poverty reduction.

Conclusion

This chapter identifies the ways in which the urgent need for climate change adaptation in urban areas provides the rationale for far stronger links between social development and the urban sector. It shows how climate change adaptation will affect "traditional" urban physical infrastructure concerns, such as housing, water, sanitation, roads, and drainage. At the same time, it identifies the crucial roles and responsibilities of individuals, households, and communities in their own adaptation processes, independent of government. Supporting such communities—and their collaboration with (and demands placed on) local institutions, such as municipal governments—will be essential if climate change adaptation is to move beyond its identification as a "technical" domain, toward recognition of the essential importance of its social dimensions. An asset adaptation framework—which assists in mapping asset vulnerability, as well as identifying interventions to strengthen, protect, and rebuild the assets and capabilities of local households and communities—is an important operational tool for ensuring that the social consequences of climate change are recognized and addressed.

References

ACHR (Asian Coalition for Housing Rights). 2005. "How Asia's Precarious Coastal Settlements Are Coping after the Tsunami." Special issue. *Housing by People in Asia* 16 (August): 1–52.

Awuor, Cynthia Brenda, Victor Ayo Orindi, and Andrew Ochieng Adwera. 2008. "Climate Change and Coastal Cities: The Case of Mombasa, Kenya." *Environment and Urbanization* 20 (1): 231–42.

Ayers, Jessica. 2009. "International Funding to Support Urban Adaptation to Climate Change." *Environment and Urbanization* 21 (1): 225–40.

Bartlett, Sheridan. 2008. "Climate Change and Urban Children: Impacts and Implications for Adaptation in Low- and Middle-Income Countries." Climate Change and Cities Discussion Paper 2, International Institute for Environment and Development, London.

Bebbington, Anthony. 1999. "Capitals and Capabilities: A Framework for Analysing Peasant Viability, Rural Livelihoods and Poverty." *World Development* 27 (12): 2021–44.

Bicknell, Jane, David Dodman, and David Satterthwaite, eds. 2009. *Adapting Cities to Climate Change: Understanding and Addressing the Development Challenges*. London: Earthscan.

Boonyabancha, Somsook. 2005. "Baan Mankong: Going to Scale with 'Slum' and Squatter Upgrading in Thailand." *Environment and Urbanization* 17 (1): 21–46.

Burra, Sundar, Sheela Patel, and Thomas Kerr. 2003. "Community-Designed, Built and Managed Toilet Blocks in Indian Cities." *Environment and Urbanization* 15 (2): 11–32.

Cabannes, Yves. 2004. "Participatory Budgeting: A Significant Contribution to Participatory Democracy." *Environment and Urbanization* 16 (1): 27–46.

Campbell, Tim. 2003. *The Quiet Revolution: Decentralization and the Rise of Political Participation in Latin American Cities*. Pittsburgh, PA: University of Pittsburgh Press.

Carter, Michael. 2007. "Learning from Asset-Based Approaches to Poverty." In *Reducing Global Poverty: The Case for Asset Accumulation*, ed. Caroline Moser, 51–61. Washington, DC: Brookings Institution Press.

d'Cruz, Celine, and David Satterthwaite. 2005. "Building Homes, Changing Official Approaches: The Work of Urban Poor Federations and Their Contributions to Meeting the Millennium Development Goals in Urban Areas." Poverty

Reduction in Urban Areas Series, Working Paper 16, International Institute for Environment and Development, London.

Douglas, Ian, Kurshid Aam, Maryanne Maghenda, Yasmin McDonnell, Louise McLean, and Jack Campbell. 2008. "Unjust Waters: Climate Change, Flooding and the Urban Poor in Africa." *Environment and Urbanization* 20 (1): 187–205.

Enarson, Elaine, and Lourdes Meyreles. 2004. "International Perspectives on Gender and Disaster: Differences and Possibilities." *International Journal of Sociology and Social Policy* 24 (10/11): 49–93.

Enarson, Elaine, Lourdes Meyreles, Marta González, Betty Hearn Morrow, Audrey Mullings, and Judith Soares. 2003. "Working with Women at Risk: Practical Guidelines for Assessing Local Disaster Risk." International Hurricane Research Center, Florida International University, Miami, FL.

Ford Foundation. 2004. "Building Assets to Reduce Poverty and Injustice." Ford Foundation, New York.

Hardoy, Jorge, Diana Mitlin, and David Satterthwaite. 2001. *Environmental Problems in an Urbanizing World: Finding Solutions for Cities in Africa, Asia and Latin America.* London: Earthscan.

Hardoy, Jorgelina, and Gustavo Pandiella. 2009. "Urban Poverty and Vulnerability to Climate Change in Latin America." *Environment and Urbanization* 21 (1): 203–24.

Hasan, Arif. 2006. "Orangi Pilot Project: The Expansion of Work beyond Orangi and the Mapping of Informal Settlements and Infrastructure." *Environment and Urbanization* 18 (2): 451–80.

Hoeppe, Peter, and Eugene N. Gurenko. 2007. "Scientific and Economic Rationales for Innovative Climate Insurance Solutions." In *Climate Change and Insurance: Disaster Risk Financing in Developing Countries,* ed. Eugene N. Gurenko, 607–20. London: Earthscan.

Huq, Saleemul, and Hannah Reid. 2007. "Community-Based Adaptation: A Briefing." International Institute for Environment and Development, London.

IPCC (Intergovernmental Panel on Climate Change). 2007. "Summary for Policy Makers." In *Climate Change 2007: Impacts, Adaptation and Vulnerability. Contribution of Working Group II to the Fourth Assessment Report of the Intergovernmental Panel on Climate Change,* ed. Martin L. Parry, Osvaldo F. Canziani, Jean Palutikof, Paul van der Linden, and Clair Hanson, 7–22. Cambridge, U.K.: Cambridge University Press.

Lavell, Allan. 1999. "Natural and Technological Disasters: Capacity Building and Human Resource Development for Disaster Management." http://www.desenredando.org/public/articulos/1999/ntd/ntd1999_mar-1-2002.pdf.

Longhurst, Richard. 1994. "Conceptual Frameworks for Linking Relief and Development." *IDS Bulletin* 25 (4): 17–23.

Manda, Mtafu A. Zeleza. 2007. "Mchenga—Urban Poor Housing Fund in Malawi." *Environment and Urbanization* 19 (2): 337–59.

McGranahan, Gordon, Deborah Balk, and Bridget Anderson. 2007. "The Rising Tide: Assessing the Risks of Climate Change and Human Settlements in Low-Elevation Coastal Zones." *Environment and Urbanization* 19 (1): 17–37.

Mitlin, Diana. 2008, "With and Beyond the State: Co-production as a Route to Political Influence, Power and Transformation for Grassroots Organizations." *Environment and Urbanization* 20 (2): 339–60.

Mitlin, Diana, and David Satterthwaite. 2007. "Strategies for Grassroots Control of International Aid." *Environment and Urbanization* 19 (2): 483–500.

Moser, Caroline O. N. 1998. "The Asset Vulnerability Framework: Reassessing Urban Poverty Reduction Strategies." *World Development* 26 (1): 1–19.

———. 2007. "Asset Accumulation Policy and Poverty Reduction." In *Reducing Global Poverty: The Case for Asset Accumulation,* ed. Caroline Moser, 83–103. Washington, DC: Brookings Institution Press.

———. 2008. "Assets and Livelihoods: A Framework for Asset-Based Social Policy." In *Assets, Livelihoods and Social Policy,* ed. Caroline Moser and Anis A. Dani, 43–82. Washington, DC: World Bank.

Moser, Caroline, and Andrew Felton. 2007. "Intergenerational Asset Accumulation and Poverty Reduction in Guayaquil, Ecuador, 1978–2004." In *Reducing Global Poverty: The Case for Asset Accumulation,* ed. Caroline Moser, 15–50. Washington, DC: Brookings Institution Press.

Moser, Caroline, Pamela Sparr, and James Pickett. 2007. "Cutting-Edge Development Issues for INGOs: Applications of an Asset Accumulation Approach." Asset Debate Paper 1, Brookings Institution, Washington, DC.

Nicolai, Susan, and Carl Triplehorn. 2003. "The Role of Education in Protecting Children in Conflict." Humanitarian Practice Network Paper 42, Overseas Development Institute, London.

ODI (Overseas Development Institute). 2004. "Disaster Risk Reduction: A Development Concern. A Scoping Study on Links between Disaster Risk Reduction, Poverty and Development." Overseas Development Institute, London.

Pelling, Mark, and Ben Wisner, eds. 2008. *Disaster Risk Reduction: Cases from Urban Africa.* London: Earthscan.

Revi, Aromar. 2008. "Furthering Pro-Poor Urban Climate Change Adaptation in Low- and Middle-Income Countries." Background paper prepared for the World Bank, International Institute for Environment and Development, London.

Roberts, Debra. 2008. "Thinking Globally, Acting Locally—Institutionalizing Climate Change at the Local Government Level in Durban, South Africa." *Environment and Urbanization* 20 (2): 521–37.

Satterthwaite, David, David Dodman, and Jane Bicknell. 2009. "Conclusions: Local Development and Adaptation." In *Adapting Cities to Climate Change: Understanding and Addressing the Development Challenges,* ed. Jane Bicknell, David Dodman, and David Satterthwaite, 359–84. London: Earthscan.

Satterthwaite, David, Saleemul Huq, Mark Pelling, Hannah Reid, and Patricia Romero Lankao. 2007. "Adapting to Climate Change in Urban Areas: The Possibilities and Constraints in Low- and Middle-Income Nations." Human Settlements Programme, Climate Change and Cities Discussion Series 1. International Institute for Environment and Development, London.

Sen, Amartya. 1997. "Editorial: Human Capital and Human Capability." *World Development* 25 (12): 1959–61.

Sisulu, Lindiwe. 2006. "Partnerships between Government and Slum/Shack Dwellers' Federations." *Environment and Urbanization* 18 (2): 401–06.

Stephens, Carolyn, Rajesh Patnaik, and Simon Lewin. 1996. "This Is My Beautiful Home: Risk Perceptions towards Flooding and Environment in Low-Income Urban Communities. A Case Study in Indore, India." London School of Hygiene and Tropical Medicine, London.

UNDP (United Nations Development Programme). 2004. "Reducing Disaster Risk: A Challenge for Development." http://www.undp.org/cpr/disred/rdr.htm.

UN-Habitat (United Nations-Habitat). 2007. *Enhancing Urban Safety and Security: Global Report on Human Settlements 2007*. London: Earthscan.

United Nations. 2008. "World Urbanization Prospects: The 2007 Revision." CD-ROM. POP/DB/WUP/Rev.2007. Department of Economic and Social Affairs, Population Division, United Nations, New York.

———. 2009. *Global Assessment Report on Disaster Risk Reduction: Risk and Poverty in a Changing Climate*. Geneva, Switzerland: United Nations.

Velásquez, Luz Stella. 1998. "Agenda 21: A Form of Joint Environmental Management in Manizales, Colombia." *Environment and Urbanization* 10 (2): 9–36.

Wamsler, Christine. 2007. "Bridging the Gaps: Stakeholder-Based Strategies for Risk Reduction and Financing for the Urban Poor." *Environment and Urbanization* 19 (1): 115–42.

Weru, Jane. 2004. "Community Federations and City Upgrading: The Work of Pamoja Trust and Muungano in Kenya." *Environment and Urbanization* 16 (1): 47–62.

Wilbanks, Tom, Patricia Romero Lankao, Manzhu Bao, Frans Berkhout, Sandy Cairncross, Jean-Paul Ceron, Manmohan Kapshe, Robert Muir-Wood, and Ricardo Zapata-Marti. 2007. "Industry, Settlement and Society." In *Climate Change 2007: Impacts, Adaptation and Vulnerability. Contribution of Working Group II to the Fourth Assessment Report of the Intergovernmental Panel on Climate Change*, ed. Martin L. Parry, Osvaldo F. Canziani, Jean Palutikof, Paul van der Linden, and Clair Hanson, 357–90. Cambridge, U.K.: Cambridge University Press.

Williams, J.R.A., Tina Hyder, and Susan Nicolai. 2005. "Save the Children's Experience: ECD in Emergencies." *Early Childhood Matters* 104: 16–21.

CHAPTER 10

Social Policies for Adaptation to Climate Change

Rasmus Heltberg, Paul Bennett Siegel, and Steen Lau Jorgensen

Integrating Adaptation with Climate Change and Development Policies

Although societies have long records of adapting to climate risks and climate changes, management of climate fluctuations continues to be costly, inadequate, and ineffective in mitigating humanitarian disasters. Climate changes are accelerating and will lead to wide-ranging shifts in climate conditions, such as temperature, precipitation, wind patterns, and extreme weather events. Developing countries are expected to see the most adverse impacts because of their geographic exposure, reliance on climate-sensitive sectors, low incomes, and weak adaptive capacity · (Cline 2007; Parry et al. 2007; Stern 2006).

How poor countries will cope with the impact of these ecological changes on their social systems remains an unanswered question. There

This chapter originated as an effort to conceptualize adaptation as social climate risk management, and it draws on some of the ideas published in Heltberg, Siegel, and Jorgensen (2009). Many people have helped us develop and refine the ideas expressed here though stimulating discussions and comments on earlier drafts, including Arun Agrawal, Harold Alderman, Catherine Arnold, Carine Clert, Rahul Malhotra, Robin Mearns, Andy Norton, Jon Padgham, Nicolas Perrin, and Tim Waites. The findings, interpretations, and conclusions expressed in this chapter are entirely those of the authors. They do not necessarily represent the view of the World Bank, its executive directors, or the countries they represent.

is growing emphasis on preparing for ongoing and future climate changes via adaptation—a process whereby societies improve their ability to manage climate risks and climate fluctuations. A common definition of "adaptation" therefore focuses on reducing risk or realizing benefits associated with climate change. Sometimes the definition is extended to include resilience to climate variability, regardless of cause, thereby framing the goal of adaptation as poverty alleviation and vulnerability reduction more than as climate management (Sabates-Wheeler, Mitchell, and Ellis 2008).

Donor agencies and developing-country governments have begun working to accelerate adaptation. However, the body of knowledge that guides the design of adaptation interventions in developing countries is limited. In particular, there is little understanding of how to prioritize adaptation investments, policies, and programs, and how to identify country-level barriers (policy, knowledge, technology) to effective adaptation. There is even less understanding of how to synchronize climate action with other social goals, such as poverty alleviation, gender balance, and empowerment. All of this raises central questions about how we conceptualize vulnerability and what role policy should play in mitigating the impacts of weather events on poor people.

This chapter discusses how to make adaptation pro-poor—which we take to mean actions that reduce poor people's vulnerability[1] to climate change (Vernon 2008). We argue that developing countries and donor agencies preparing for ongoing and future climate changes could usefully focus on actions that are "no regrets" and are multisectoral and multilevel, that improve the management of current climate variability, and that integrate adaptation with general development. Social policy and social protection, already concerned with vulnerability reduction, can promote pro-poor adaptation to climate change through a range of social programs that build the resilience of the poor, either directly or indirectly. Policy measures to this end might include community-driven adaptation, safety nets for climate risks and natural disasters, livelihoods programs, microfinance, and index insurance.

In the succeeding sections of the chapter, we survey the poverty implications of current and future climate variability; set out some principles for adaptation; discuss the role of local, national, and global efforts; and identify governance challenges to adaptation. We then focus on the design of social policy interventions for pro-poor adaptation before offering conclusions.

Policy Responses to Climate Volatility

Climate changes impact poverty through long-term changes in ecosystems and livelihoods and through greater volatility in climate conditions. The households most vulnerable to climate change are those whose assets and livelihoods are directly dependent on climate patterns and who have weak risk management capacity. The most affected livelihoods will be in many of the natural resource–intensive sectors, such as cropping, fishing, livestock, forestry, and firewood collection. Within households, impacts are likely to fall disproportionately on particularly vulnerable individuals, such as children, women, the elderly, and persons with disabilities. Although no credible estimates exist, there is consensus that, in most developing regions, climate changes could result in lower mean returns to assets and livelihoods, lower expected levels of well-being and higher poverty, more fluctuations in well-being, and increased difficulty maintaining and rebuilding assets and livelihoods in the wake of natural disasters.

Predictions of likely climate change indicate developing countries will see increased weather volatility and natural disasters. Given recent experience with managing climate fluctuations, these predictions do not bode well for the world's poor. In many parts of Africa, current rainfall and temperature fluctuations already cause unpredictable agricultural production to drive food security cycles that oscillate between times of food surplus and of famine. Among the predicted consequences of climate change are increased frequency and severity of disasters, both slow and of rapid onset. Natural disasters add to the myriad of problems facing the poor and vulnerable. Poor households suffer from limited quantity and quality of assets and volatile asset returns. The situation is worsened by social differentiation and exclusion that frequently transcend generations and lead to entrenched discrimination, institutionalized inequalities, and limited access to jobs and community resources (World Bank 2005; Leach, Mearns, and Scoones 1999). Inadequately managed disaster risk exacerbating the effects of natural catastrophes can wipe out years of development progress with severe loss of life and the destruction of livelihoods. As Sen (1981) has recognized, famines are human-made disasters—a combination of diminished food production resulting from climate risks and subsequent counterproductive human responses to food scarcity.

Climate events often have long-term implications because of the irreversible loss of human and physical capital. A longitudinal study in Zimbabwe followed children who were less than 2 years old (the age at which children are most susceptible to malnutrition) when a severe drought hit in the early 1980s. Those who survived the famine were found to be stunted. Their stunting translated into lower school achievement, inferior adult health, and an estimated 14 percent reduction in lifetime earnings (Alderman, Hoddinott, and Kinsey 2006). Studies of Hurricane Mitch in Honduras showed that the hurricane exacerbated asset inequalities because the poor lost a greater share of assets in the disaster and recovered at a slower rate than did those who were not poor. And analysis of rural Ethiopian households hit by drought showed that, although better-off households could sell livestock to finance consumption, the poorer households often tried to hold on to their livestock at the expense of food consumption to preserve their options for rebuilding herds. Those poor households that exited the drought with few or no assets faced great difficulty rebuilding their herds (Carter et al. 2007).

Current polices are having limited success in mitigating such effects. Existing arrangements for managing climate and other risks offer poor households limited protection from adverse impacts. The fluctuations in consumption and in human and physical assets that result from climate and other shocks have adverse consequences for household well-being and for economic growth (Dercon 2004). Short-run impacts on households include reductions in the quality and quantity of food, health, and education, as well as longer hours worked. Long-run impacts include destitution, landlessness, asset loss, irreversible malnutrition, child labor, and withdrawal from schooling. Recovery from shocks often is slow and incomplete. Survey data have shown that the poor struggle to repay debts and rebuild assets (Heltberg and Lund 2009). Many of the very poor are failed almost entirely by both government and market-based social protection and risk management instruments; and they are forced to rely largely on informal coping responses, such as self-insurance, asset decumulation, and assistance from informal networks.

Large-scale natural disasters sometimes trigger government and donor assistance, although it tends to be too little, too late, and in the form of food rather than cash. Although critical and potentially lifesaving, such support is often ad hoc and not part of any long-term strategy to protect household livelihoods from shocks. If international support for humanitarian assistance does materialize, often it is only after a dire situation has reached the brink of utter collapse (Barrett and Maxwell 2005). For example, in the Niger famine of 2005–06, early-warning systems accurately predicted a looming disaster; however, it was a long time before assistance arrived in the affected areas. During that time, households had to choose between distress sales of productive assets and destruction of human assets (for example, malnutrition and removing children from school). Such decisions have long-term poverty implications. There also is concern that donor-funded relief may undermine countries' incentives for crafting, and paying for, national social protection systems and weather-based insurance.

Internalization of risk triggers additional efficiency losses. For instance, households anticipating uninsured risk often engage in low-risk, low-return activities—such as maintaining short-term asset liquidity in lieu of investment and long-term returns. A study in southern India found that in the presence of high risk, poor farmers reaped lower returns to assets than did the better-off farmers, whereas the reverse was true in low-risk settings (Rosenzweig and Binswanger 1993). In short, high risk and the absence of effective risk management instruments conspire to constrain asset growth and the escape from poverty. The adverse impacts on asset growth of the ex ante behavioral responses to risk may outweigh even the ex post impacts (Elbers, Gunning, and Kinsey 2007).

In sum, current social responses to ongoing climate volatility have failed to offer effective protection to the poor. If climatologists are correct in their predictions that weather patterns will become increasingly volatile because of climate change, then development professionals and the leaders of poor countries must begin to think more seriously about what strategies can be used to mitigate the effects of climate volatility on the livelihoods of the poor.

Some Principles for Adaptation

Adaptation to ongoing and future climate changes means many different things, depending on country, sector, and climate risk. The subject of adaptation is complicated further by the uncertainties surrounding climate predictions and the pathways of potential impact. Given the current state of knowledge, development professionals are far from being able to draw up definitive best practices for adaptation—let alone a "blueprint." However, the following principles for good adaptation have been proposed and seem to make sense as guideposts:

- Identify "no-regrets" instruments—investments and policies with high payoffs under the current climate and in a future with a different and more volatile climate.

- Improve management of current climate variability to prepare for worse weather ahead. Help households replace unproductive, asset-degrading coping strategies (such as withdrawing children from school, delaying health care, and selling long-term assets to meet immediate needs) with ex ante mechanisms that anticipate, plan, and act against the negative impacts of risks.
- Consider risk and responses in an integrated, multisectoral fashion rather than for each risk and each sector alone. For example, water resources management will be of growing importance for many sectors, such as agriculture, hydropower, and urban development.
- Adaptation is good development, and good development is adaptation. The financing and planning for adaptation must be integrated with general development finance and planning.
- Adapt at many levels. Efforts by households, communities, and nations to respond to climate change must be complemented by international responses based on the principles of global burden sharing and social justice.
- Prepare for long-term, continued engagement on adaptation with communities and countries. Climate change is not an issue that will be resolved in the near term.

Local, National, and Global Efforts

Although climate change is a global process, the way it manifests varies by locality. Likewise, responses to climate change are often local: adaptation, whether planned or spontaneous, takes place largely through a myriad of decisions by households, communities, and local organizations.[2]

Households and communities do their best to adapt to perceived climate changes, even in the absence of facilitating government policy and sometimes despite policy or regulatory constraints. Producers have private incentives to explore investment opportunities to adjust assets, technologies, and livelihoods to changing climates. However, these incentives are not always sufficient. Information, technologies, and financing are also needed to help producers benefit from opportunities and overcome market imperfections. In addition, the poorest and most disadvantaged will need further help in overcoming entrenched barriers that exclude them from full participation in social and economic life.

Local social networks sometimes substitute for state-led actions, as when communities take over coastal zone or forest resource management (Adger 2003) or when informal transfers are used to help households survive shocks in the absence of government and market-based social protection instruments. Such community-driven adaptation often is driven by necessity—local people forced to protect their livelihoods when nobody else does it for them—but it needs and deserves external support. In the absence of robust support, the scale and covariate nature of many of the risks associated with climate change could overwhelm community coping mechanisms. External support should recognize the strengths of communities and seek to exploit those strengths for effective adaptation. With their extensive local knowledge of people and ecosystems, communities must be an integral part of planning and implementing adaptation. Community-led interventions ought to form part of country adaptation programs, and affected communities should be consulted in the planning process.

It is unfortunate that some current approaches to adaptation planning and financing may bypass local institutions. The current push to formulate national adaptation plans of action seems to have missed the opportunity to propose adaptation projects for community- or local-level public, private, or civic institutions. According to Agrawal (2008; see also chapter 7 of this volume), only 20 percent of projects described in the national adaptation plan-of-action documents incorporate local institutions as the focus of adaptation projects; even fewer identify local institutions as agents or partners in facilitating adaptation (Agrawal 2008, pp. 42–44). As more external finance for adaptation becomes available, it will be important to identify a robust mechanism for channeling support to local initiatives.

Unless assisted, some of the most adversely affected countries stand at risk of dramatic upheavals with potentially serious regional and global spillovers. For example, major declines of food production in entire regions of Africa, advanced desertification of large regions, or the drying up of rivers that supply regions and cities with water all have international implications. Such regional disasters could overwhelm entire countries. There also are issues of equity and social justice because the poor stand to lose the most from a problem they have done the least to create. Equity and fairness therefore dictate that a substantial part of the burden for adaptation ought to come from the international community.

Governance Challenges

Some of the most vulnerable households are in communities and countries that have the weakest institutional capacity and the fewest resources to respond. Addressing this challenge will require responsive and accountable government institutions. However, such institutions cannot be created simply using external financing. As in other areas of development, the key barrier to progress lies in overcoming the governance challenges to ensure the risks facing poor people are acknowledged and addressed effectively. Interventions are unlikely to meet with success and attract sustained donor funding unless government institutions can demonstrate accountability and responsiveness.

International institutions, therefore, will have to find ways to provide not only resources but also incentives and information for adaptive actions that are responsive to the poor. It is here that social protection enters. As discussed more extensively below, risk-responsive safety net programs and adaptation support through social funds and community-driven development projects should be considered. There is also work to be done on identifying barriers to adaptation rooted in counterproductive policies and regulations. For example, lack of clear and enforced property rights to land undermines incentives to invest in land improvements and irrigation. These investments are needed to address the projected declines in agricultural productivity and the increased volatility of precipitation.

International sharing of the burdens of adapting to climate change will need to go beyond development assistance; it will have to include labor and migration policies, water sharing, food trade, financial markets and insurance systems, and possibly even peacekeeping when degradation and resource scarcities caused by climate change trigger violent conflict. Climate change, in other words, is a truly global social and environmental issue with spillovers for a range of contested international issues.

Social Policy and Social Protection for Adaptation

So far, many interventions for adaptation have focused on the role of specific sectors—such as energy, communications, or water—and on the "climate proofing" of infrastructure projects. There has also been some discussion in the agricultural sector about how to modify crops in the context of climate change. However, more attention must be paid to social issues, such as indirect risks, household vulnerability, disaster risk management, and inclusion of poor and vulnerable people. A robust social policy response, rooted in an understanding of the risks associated with climate change and climate vulnerability facing the poor, is needed to make climate action more pro-poor. There also is a role for social policy to empower the poor and help them develop the voice and political assets needed to claim access to risk management instruments. For that reason, Stern (2008) considers social protection among the priority sectors for adaptation in developing countries. With increases in concessional assistance for climate change and adaptation quite likely, there could be potential for increased external support for the social sectors.

Within the context of climate change adaptation, social policy can play a unique role because of its ability to create a policy space where the trade-offs between ecological concerns and poverty can be negotiated. Interventions addressing climate change sometimes are at odds with poverty and other development objectives. For example, efforts to expand forest carbon finance have met with criticism from some indigenous communities who fear it could undermine further their often-tenuous property rights to ancestral lands. Another example is the expansion of biofuel production causing food prices to spike, with severe impacts on food-insecure households in many countries. In contrast, social policy approaches to adaptation create synergies between climate action and poverty alleviation. What is key is understanding the risks associated with climate change and climate vulnerability facing poor and vulnerable people, and designing instruments that help people manage these risks.

Experience with social policy approaches that are responsive to climate risks is building. Despite the perennial problem of dealing with climate shocks, coverage of programs and instruments helping poor and vulnerable people manage climate risks remains very low; pro-poor adaptation should aim to change that situation. Interventions that might be successful in meeting those needs should focus on social funds, social safety nets for natural disasters, livelihoods, microfinance, and index insurance (see, for example, the articles in Tanner and Mitchell 2008; IDS 2007; and Yamin, Rahman, and Huq 2005).

Social Funds for Community-Based Adaptation

Social funds are semiautonomous institutions created to channel external support to communities. Social funds and community-driven development programs support small projects in a number of sectors (for example, infrastructure, social services, microenterprise development, microfinance, forestry, and ecosystem management) that have been identified by communities and presented to the social fund for financing. Social funds and community-driven development programs allow poor people and communities to become actively involved in their own development.

Social funds can support pro-poor adaptation by scaling up their work in sectors relevant for creating resilience, such as ecosystem management and restoration, water supply and sanitation, community forestry, coastal zone management, disaster preparedness, and postdisaster assistance. Even in countries with weak capacity, the international community could use social funds and community-driven development to channel external finance to small-scale community adaptation projects at scale. This could also encompass community-driven investments in preserving woodlands and forests to attract carbon finance.

Social Safety Nets for Coping with Natural Disasters and Climate Shocks

Because traditional safety net programs have targeted the chronically poor, their use in mitigating the effects of climate shocks remains relatively underexplored. However, there is growing interest in applying safety nets to help avoid postdisaster famine and in helping affected households and communities protect and rebuild their assets. Conditional and unconditional cash transfers, workfare programs, and in-kind transfers are some of the available instruments. Large-scale cash support to affected households has been an important and well-performing part of the disaster response in recent major natural disasters in South Asia and Turkey. It makes sense to prepare for better design and for swifter and more equitable and consistent deployment after weather shocks. The key preparatory step is to build country capacity to deliver cash transfers or execute public works after natural disasters. The same capacity can be used to cope with food, fuel, conflict, and financial shocks. Countries and donors should work this into their disaster preparedness strategies (Heltberg 2007; Vakis 2006).

It is useful but uncommon for existing safety net programs to have contingency arrangements in place for scaling up (Grosh et al. 2008). For such programs to be effective, they must be fully operational prior to natural disasters; and must maintain flexibility in their targeting, financing, and implementation arrangements (Alderman and Haque 2006; de Janvry et al. 2006). Such programs allow countries to provide immediate relief and rehabilitation assistance to disaster victims. For example, Bangladesh has built capacity for rapid transfers of in-kind or cash payments to disaster victims. As part of an integrated approach to disaster risk management that also includes large-scale private rice imports, these transfers have helped reduce the mortality from natural disasters and improve disaster recovery. In Honduras, an existing social fund was able to scale up labor-intensive community projects after Hurricane Mitch. Within a few months, these projects created a large amount of temporary employment in communities where infrastructure had been disrupted by the hurricane (Grosh et al. 2008).

There is an additional benefit to assuring households in advance that they will receive disaster benefits under specified conditions. As a substitute for insurance, such assurances would allow households to make livelihood decisions with higher risks and higher expected returns. Programs and pilot projects that are under way in Ethiopia, India, Mexico, and Mongolia offer interesting innovations in weather risk management by combining insurance and safety net or social insurance approaches. These programs use weather indexes as triggers to mobilize safety net transfers and payouts to farmers or herders. Ethiopia's Productive Safety Nets Program may be the best example of this approach. The program offers a combination of cash transfers and public workfare to approximately 6 million chronically food-insecure people. The aims of the program are to reduce household vulnerability, improve household and community resilience to shocks, and break the country's dependence on food aid. The program has developed a mechanism based on rainfall indexes for temporary expansion into drought-affected areas threatened with food shortages.

Livelihoods Programs

Access to assets and employment is vital for building the resilience of the poor. As the productivity of many natural resource–based livelihoods declines, peoples' transition into new livelihoods—often in new sectors and in urban areas—may need temporary support. Social protection can support such livelihood diversification through employment generation, asset transfers and asset building, livestock restocking, seed transfers, training and skills development, microfinance initiatives, more orderly migration, and access to remittances that is safe and easy. Employment generation programs, such as Maharastra's Employment Guarantee Scheme, have shown that it is possible to transfer and stabilize incomes while building valuable community assets (Grosh et al. 2008). Programs also may focus on building the assets of the poor and protecting the returns to those assets through access to markets and protection of rights (Davies et al. 2009). In Nicaragua, the World Bank and bilateral donors support an innovative pilot program that combines conditional cash transfers with additional transfers aimed at increasing the income-generating capacity of poor rural households exposed to weather risk. Such efforts can reduce poverty and improve resilience simultaneously.

Microfinance

Access to financial products and services remains an underserved area that is important in helping poor people smooth consumption and manage risk. A large proportion of low-income people is excluded from the banking system and forced to use less secure and more inflexible methods of payment, such as cash, informal borrowing, and informal money transfers. Poor people face many barriers to financial access—distance from services, the inability to produce formal documents, and prohibitive costs. Across sub-Saharan Africa, for example, only 20 percent of households have accounts with financial institutions (World Bank 2008, p. 35).

Microfinance can help bridge this gap and support adaptation through livelihood support and risk management instruments (Hammill, Matthew, and McCarter 2008). Microfinance is the delivery of loans, savings, insurance, and other financial services to low-income groups so they can engage in productive activities, build assets, and protect themselves against risk. Microfinance often does not cover the very poorest who are considered unbankable, but it serves a slightly more stable low-income segment that has been ignored by traditional banks. In Bangladesh, however, the large nongovernmental organization BRAC (Bangladesh Rural Advancement Committee) has pioneered approaches to sequencing safety net support, skills building, and microfinance in a program that aims to "graduate" the poorest people into microfinance clients.

Lending is the best-known component of microfinance. Microloans most often are given for productive purposes—to purchase a small asset or finance working capital. These loans help low-income people start a small business and reduce their reliance on moneylenders and other informal sources of credit. Microloans are not directly designed for risk coping, which would require lending to expand in times of shocks. Instead, they create resilience indirectly, to the extent they are successful at helping people grow and diversify their incomes and assets and reduce their vulnerability. Microsavings complement microloans by giving low-income groups access to safe and cheap savings instruments.

There is increasing interest in microinsurance as a direct risk management instrument. Formal insurance is often superior to informal risk management relying on low-risk, low-return strategies and on social connections that may or may not offer help in times of need. Microinsurance aims to increase the outreach and coverage of formal insurance across lower income tiers that otherwise would not be covered by more traditional insurance companies. It does so by tailoring insurance products to the needs and purchasing power of low-income people. Common microinsurance products include health, life, and index-based (often linked to rainfall) insurance. Whereas microinsurance can be offered by anyone, including nongovernmental organizations and the private sector, microfinance institutions have shown the biggest interest and have begun offering life insurance as a way to insure their outstanding loans (Alderman and Haque 2007). Social funds also have shown interest in microinsurance. Although their lack of actuarial expertise can be problematic, the existing relationship between social funds and potential clients can help establish the trust that is critical to the uptake of any insurance product (Maleika and Kuriakose 2008).

Weather-Based Index Insurance and Other Conditional Financial Instruments

Emerging lessons with index insurance at the household, local, or national levels show scope for insuring against low-frequency and high-cost weather events. Weather-based index insurance sometimes can substitute for traditional crop insurance, which has had little success in developing countries because of the high administrative costs to verify claims. Weather-based index insurance uses objectively defined trigger events (for example, rainfall or soil moisture) in an area to set contingent damage payments according to an index. Buyers may be farmers or local and national governments. Contracts and indemnity payments are the same for all buyers, per unit of insurance; there is no use of field- or household-specific damage and loss data. In contrast to microinsurance, weather-based insurance can be offered to countries or regions—as is the case with hurricane insurance, for example. Index insurance discourages moral hazard and cheating, avoids adverse selection problems, and lowers transaction costs. It also makes the insurance instrument accessible to the broader rural population (Skees et al. 2002). However, index insurance weakens the correlation between losses and payouts, a problem known as "basis risk"—an insured party may suffer a loss, but not receive a payout. Index insurance still presents technical challenges, such as data availability, and may not be easily affordable or in high demand in many countries.

Index insurance is not a panacea (Alderman and Haque 2007). It may not be appropriate for slow-onset climate impacts; and preventing losses is sometimes more cost-effective than providing loss-based insurance. Furthermore, many low-income countries lack insurance markets and may not find insurance easily affordable. It also may not be desirable for some developing countries to take out insurance if indemnities crowd out concessional emergency funding. Therefore, weather-based insurance cannot stand alone. Many humanitarian crises are caused by factors other than climate variability—by conflict, poor governance, lack of infrastructure, political instability, and macroeconomic shocks. Safety net and emergency response policies thus should not be tied exclusively to index instruments.

Conclusion

There is much uncertainty about the socioeconomic implications of climate change and how best to design adaptation. Unless societies adapt, risks associated with climate change could cause large financial losses, increased vulnerability, and more frequent humanitarian disasters. Developing countries and donor agencies, therefore, should do more to prepare for ongoing and future climate changes, focusing on no-regrets actions that are multisectoral and multilevel and that improve the management of current climate variability. In planning and financing, adaptation should be integrated with general development. Social scientists and development practitioners must step up to this challenge with the aim of promoting adaptation that is sustainable, pro-poor, and on a scale commensurate with the challenges. It is crucial to understand the risks associated with climate change and climate volatility that poor and vulnerable people face, and to design and scale up instruments for managing those risks.

Although there is a long way to go, proven social policy frameworks and instruments for reducing vulnerability and involving communities in development will be valuable for this endeavor. As mentioned, interventions to take forward include social funds that operate on the principles of community-driven development, safety nets that better respond to climate risks and natural disasters, livelihoods programs, microfinance, and index insurance.

Notes

1. We define "household vulnerability" as the expectation of falling below benchmark levels of well-being (for example, below the poverty line) if a risky event occurs. An individual or household is vulnerable to risks associated with climate change if these risks will result in a loss that pushes the household below the well-being benchmark. In our definition, vulnerability depends on the characteristics of the risks, exposure and sensitivity to the risks, expected impacts and losses, and risk management capacity.

2. Elsewhere, we have explored how the social risk management framework can be used to analyze the choices between risk management at the household, local, and national levels, and between ex ante anticipatory instruments and ex post coping responses (Heltberg, Siegel, and Jorgensen 2009).

References

Adger, W. Neil. 2003. "Social Capital, Collective Action, and Adaptation to Climate Change." *Economic Geography* 79 (4): 387–404.

Agrawal, Arun. 2008. "The Role of Local Institutions in Adaptation to Climate Change." Paper prepared for the World Bank Social Dimensions of Climate Change Workshop, Washington, DC, March 5–6.

Alderman, Harold, and Trina Haque. 2006. "Countercyclical Safety Nets for the Poor and Vulnerable." *Food Policy* 31 (4): 372–83.

———. 2007. *Insurance Against Covariate Shocks: The Role of Index-Based Insurance in Social Protection in Low-Income Countries of Africa.* Africa Human Development Series, World Bank Working Paper 95. Washington, DC: World Bank.

Alderman, Harold, John Hoddinott, and Bill Kinsey. 2006. "Long Term Consequences of Early Childhood Malnutrition." *Oxford Economic Papers* 58 (3): 450–74.

Barrett, Christopher B., and Daniel G. Maxwell. 2005. *Food Aid After Fifty Years: Recasting Its Role.* London: Routledge.

Carter, Michael R., Peter D. Little, Tewodaj Mogues, and Workneh Negatu. 2007. "Poverty Traps and Natural Disasters in Ethiopia and Honduras." *World Development* 35 (5): 835–56.

Cline, William R. 2007. *Global Warming and Agriculture: Impact Estimates by Country.* Washington, DC: Center for Global Development, Peterson Institute for International Economics.

Davies, Mark, Bruce Guenther, Jennifer Leavy, Tom Mitchell, and Thomas Tanner. 2009. *Climate Change Adaptation, Disaster Risk Reduction and Social*

Protection: Complementary Roles in Agriculture and Rural Growth? Brighton, U.K.: Institute of Development Studies.

de Janvry, Alain, Elisabeth Sadoulet, Pantelis Solomon, and Renos Vakis. 2006. *Uninsured Risk and Asset Protection: Can Conditional Cash Transfer Programs Serve as Safety Nets?* Social Protection Discussion Paper 604. Washington, DC: World Bank.

Dercon, Stefan, ed. 2004. *Insurance against Poverty.* Oxford, U.K.: Oxford University Press.

Elbers, Chris, Jan Willem Gunning, and Bill Kinsey. 2007. "Growth and Risk: Methodology and Micro Evidence." *World Bank Economic Review* 21 (1): 1–20.

Grosh, Margaret, Carlo del Ninno, Emil Tesliuc, and Azedine Ouerghi. 2008. *For Protection and Promotion: The Design and Implementation of Effective Safety Nets.* Washington, DC: World Bank.

Hammill, Anne, Richard Matthew, and Elissa McCarter. 2008. "Microfinance and Climate Change Adaptation." *IDS Bulletin* 39 (4): 113–22.

Heltberg, Rasmus. 2007. "Helping South Asia Cope Better with Natural Disasters: The Role of Social Protection." *Development Policy Review* 25 (6): 681–98.

Heltberg, Rasmus, and Niels Lund. 2009. "Shocks, Coping, and Outcomes for Pakistan's Poor: Health Risks Predominate." *Journal of Development Studies* 45 (6): 889–910.

Heltberg, Rasmus, Paul Siegel, and Steen Lau Jorgensen. 2009. "Addressing Human Vulnerability to Climate Change: Toward a 'No Regrets' Approach." *Global Environmental Change* 19: 89–99.

IDS (Institute of Development Studies). 2007. *IDS in Focus: Climate Change Adaptation.* Issue 2, November. Brighton, U.K.: IDS.

Leach, Melissa, Robin Mearns, and Ian Scoones. 1999. "Environmental Entitlements: Dynamics and Institutions in Community-Based Natural Resource Management." *World Development* 27 (2): 225–47.

Maleika, Marc, and Anne T. Kuriakose. 2008. "Microinsurance: Extending Pro-Poor Risk Management through the Social Fund Platform." *Social Funds Innovations Notes* 5 (2): 1–8.

Parry, Martin, Osvaldo F. Canziani, Jean Palutikof, Paul van der Linden, and Clair Hanson, eds. 2007. *Climate Change 2007: Impacts, Adaptation and Vulnerability. Contribution of Working Group II to the Fourth Assessment Report of the Intergovernmental Panel on Climate Change.* Cambridge, U.K.: Cambridge University Press.

Rosenzweig, Mark R., and Hans P. Binswanger. 1993. "Wealth, Weather Risk and the Composition and Profitability of Agricultural Investments." *Economic Journal* 103 (416): 56–78.

Sabates-Wheeler, Rachel, Tom Mitchell, and Frank Ellis. 2008. "Avoiding Repetition: Time for CBA to Engage with the Livelihoods Literature?" *IDS Bulletin* 39 (4): 53–59.

Sen, Amartya. 1981. *Poverty and Famines: An Essay on Entitlement and Deprivation.* Oxford, U.K.: Oxford University Press.

Skees, Jerry R., Panos Varangis, Donald F. Larson, and Paul B. Siegel. 2002. "Can Financial Markets Be Tapped to Help Poor People Cope with Weather Risks?" Policy Research Working Paper 2812, World Bank, Washington, DC.

Stern, Nicholas. 2006. *The Economics of Climate Change: The Stern Review.* Cambridge, U.K.: Cambridge University Press.

———. 2008. "Key Elements of a Global Deal on Climate Change." London School of Economics and Political Science, London.

Tanner, Thomas M., and Tom Mitchell. 2008. "Introduction: Building the Case for Pro-Poor Adaptation." *IDS Bulletin* 39 (4): 1–5.

Vakis, Renos. 2006. *Complementing Natural Disasters Management: The Role of Social Protection.* Social Protection Discussion Paper 543. Washington, DC: World Bank.

Vernon, Tamsin. 2008. "The Economic Case for Pro-Poor Adaptation: What Do We Know?" *IDS Bulletin* 39 (4): 32–41.

World Bank. 2005. *World Development Report 2006: Equity and Development.* Washington, DC: World Bank.

———. 2008. *Finance for All? Policies and Pitfalls in Expanding Access.* World Bank Policy Research Report. Washington, DC: World Bank.

Yamin, Farhana, Atiq Rahman, and Saleemul Huq. 2005. "Vulnerability, Adaptation and Climate Disasters: A Conceptual Overview." *IDS Bulletin* 36 (4): 1–14.

CHAPTER 11

Seeing People through the Trees and the Carbon: Mitigating and Adapting to Climate Change without Undermining Rights and Livelihoods

Andy White, Jeffrey Hatcher, Arvind Khare, Megan Liddle, Augusta Molnar, and William D. Sunderlin

In the next few decades, the world will face an unprecedented sequence of challenges. Global markets and political structures are shifting. Global climate change is already beginning to alter weather events, seasonal patterns of precipitation, temperature, and wildlife distribution. In the midst of these

The individuals who have authored and contributed to this chapter are part of the Rights and Resources Initiative (RRI). RRI is a global coalition working to advance forest tenure, policy, and market reforms, primarily in developing countries. Our mission is to promote pro-poor forest policy and market reforms that will increase household and community ownership, control, and benefits from forests and trees. Partners in the RRI coalition are the Coordinating Association of Indigenous and Community Agroforestry in Central America, the Center for International Forestry Research, Civic Response, the Federation of Community Forest Users of Nepal, Forest Peoples Programme, the Foundation for People and Community Development, Forest Trends, Intercooperation, the International Union for Conservation of Nature, Regional Community Forestry Training Center for Asia and the Pacific, and the World Agroforestry Centre. For more information, visit http://www.rightsandresources.org. This chapter is based on analysis from the report *Seeing People Through the Trees: Scaling Up Efforts to Advance Rights and Address Poverty, Conflict and Climate Change* (RRI 2008), and on discussion from the October 2008 international conference on Rights, Forests and Climate Change, organized by Rainforest Foundation Norway and RRI (http://www.rightsandclimate.org). The analysis includes contributions from Liz Alden Wily, Jürgen Blaser, Intu Boedhihartono, Sarah Byrne, Doris Capistrano, Marcus Colchester, Bob Fisher, Brooke Kennedy, Ruben de Koning, Stewart Maginnis, Jeffrey McNeely, Sten Nilsson, Carmenza Robledo, Don Roberts, Jeffrey Sayer, Kaspar Schmidt, Gill Shepherd, and Yurdi Yasmi.

challenges, the global development agenda—which only recently peaked with the identification of the Millennium Development Goals (MDGs)—has lost ground to the more politically pressing issues of security: food security, energy security, political security, and ecological security.

Forests are central to understanding and addressing many of these challenges. More than 18 percent of global carbon dioxide emissions stem from deforestation, forest degradation, and land-use change (Stern 2006, p. 537). Global market demands for commodities, including bioenergy, are increasing pressure on forestlands and forest peoples. So are the emerging markets for forest carbon and political responses to security challenges. More than ever, the markets and politics of forests and forest peoples are interlinked with those of the global community. Our fates are intertwined: our consumption affects their lands; our carbon dioxide emissions affect their forests (Menzies 2007).

Unfortunately, despite some 50 years of development assistance, the forest frontier in developing countries continues to recede. The conventional conservation and development models promoted by development organizations have proved ineffective, by and large, in establishing sustained conservation, development, or economic growth in forest areas. There is high risk that with climate change, attempts to use forests to mitigate climate change, and oncoming market transitions, millions of people will be pushed farther into poverty and conflict; and that distinct cultures will be pushed to extinction. How tensions over forests play out in coming decades will influence the severity of climate change, the course of wars and civil conflicts, and the health of the world that our descendants will inherit.

Given this history, there is good reason to be concerned that the new funding flowing into forest areas from climate change initiatives succeed in addressing these key failures of the past. Crucial to any success will be establishing the sound institutional footing needed for equitable social and economic development, as well as fair markets—including the recognition and clarification of property rights; establishment of accountability mechanisms; development of transparency in government decisions; and empowerment of local people to fully participate as citizens in the decisions that affect their rights, interests, and livelihoods.

We argue that, with robust and proactive steps, climate change and the global response can be converted from a major threat to a major opportunity, not only to reduce emissions from forests, but also to advance governance and development in forest areas. To ensure that investments for climate change mitigation and adaptation in forest areas are effective—and, at a minimum, do not undermine local rights and livelihoods—our experience suggests that policy agreements and investments must work to ensure the establishment of four mutually reinforcing and self-correcting foundations:

1. recognizing and strengthening local land and resource rights and governance;
2. monitoring more than carbon;
3. independent advising and auditing of mitigation and adaptation mechanisms at both national and global levels;
4. paying the right people by prioritizing investments in the stewards of forests and trees: indigenous peoples, forest communities, and household owners.

The Problem

The 1972 United Nations Conference on the Human Environment, held in Stockholm, Sweden, was one of the first international forums to recognize the link between rights, well-being, and the environment. The Stockholm Declaration pledged to protect fundamental rights to freedom, equality, and an adequate standard of living and to safeguard the environment. In 2000, global leaders met again to set the MDGs, pledging to halve poverty and make substantial progress on other social and environmental goals by 2015. Nevertheless, 37 years after the Stockholm conference and with just 6 years to go before we reach the target date of the MDGs, the gap between aspirations and achievement is still wide. In many forest areas, the gap never closed.

Today's national and global insecurity is often driven by the same underlying problems that gave rise to the Stockholm conference and the MDGs: the inadequate recognition of human, civil, and political rights; the political and economic marginalization of rural and forest communities; widespread rural poverty; and weak and unrepresentative governing institutions.

Many forest communities, particularly in developing countries, are chronically poor and poorly governed by the state. They suffer disproportionately from conflicts, humanitarian crises, and corruption, which often then spread nationally and internationally. The property rights of forest communities are widely unrecognized; and the human, civil, and political rights of indigenous peoples, women, and other marginalized groups in forest areas are frequently limited (Colchester 2008; Sunderlin 2007; FAO 2006). More than 30 forested countries have experienced widespread violent conflict over the last 20 years, much of it caused by ethnic tension and the inequitable distribution of resources (Kaimowitz 2005, p. 5). Approximately two thirds of all violent conflicts in the world are driven by contested claims over land (Alden Wily 2008a, p. 4).

It is also clear that whereas recognition and clarification of land and resource rights are essential for enabling development as well as justice, legal reforms alone are insufficient to ensure that local people can protect, develop, and benefit from their assets. Rights reform in the forest sector can achieve the desired potential only with prior or concurrent action on broader governance issues that underpin the absence or weakness of rights at the local level. Such action includes attention to regulatory reform (forest regulations tend to favor the interests of large enterprises); market reform (to ensure that small producers of forest products have equal market opportunities); judicial reform (forest dwellers need a functioning judicial system and conflict resolution mechanisms to defend their rights); stronger public forest services (forest tenure reform often requires gazetting and demarcation of property boundaries, and that requires a sufficient budget and training); enforcement of laws against forest crimes (forest dwellers frequently fall victim to illegal appropriations of land and resources that are not prosecuted); and support for the emergence of small- and medium-scale forest enterprises. Engaging in these reforms can be an uphill battle because they challenge the status quo and vested interests, but measurable progress on forest tenure reform in recent years shows that such reforms are possible.

Many in the development community now realize that recognizing and securing land rights, strengthening civil rights, and introducing more democratic governance systems in forest areas are critical actions—not just for moral reasons, but also to achieve social, economic, and environmental goals. Fair and secure rights to natural resources, particularly land, are fundamental building blocks in any viable strategy for dealing with climate change and strengthening local and systemic resilience against future shocks. Moreover, recognizing and strengthening these rights will be key to addressing climate change mitigation and adaptation, while promoting poverty alleviation and well-being, good governance, and equitable economic growth.

The Urgency and Risks

The urgency of redressing the balance in favor of local development, rights, and resilience is greater than ever before. The dramatic shifts under way in markets, politics, and the planet's climate create new and tremendous challenges for achieving peace and prosperity in forest areas.

In recent years, the growth of the global economy and the growing demand for food, basic commodities, and energy have increased the pressure on forest peoples, who increasingly must compete for a diminishing amount of available land. The global financial crisis has reduced this pressure somewhat, but the lull is likely to be temporary. Local populations are growing, too, increasing landlessness, migration, and local pressure for the privatization of land held in common. Climate change is affecting the ecology and ranges of the flora and fauna on which forest peoples depend, and undermining livelihoods. Moreover, some of the proposed approaches to reducing carbon dioxide emissions from forests threaten to criminalize traditional land use (such as shifting cultivation), thus exacerbating existing tensions and eliminating local livelihoods.

In the past decade, the amount of forest designated as public parks and protected areas has almost doubled, most often at the expense of the people who inhabit or depend on these areas (West, Igoe, and Brockington 2006). The relative weakness of local organizations and a lack of safeguards and accountability facilitate what has been called the "great green land grab," in which private investors and conservationists rush to lock up natural forest areas before they can be converted to other land uses (Vidal 2008). In sum, this new set of pressures raises the risk not only of increased poverty, social exclusion, and civil conflict among forest peoples; but also the risk of increased carbon dioxide emissions from continued or increased deforestation and forest degradation.

Forest Areas and Development:
Current Status and Lessons from History

Although development aid and cooperation can claim successes over the past five decades, few of these successes have been in forest areas. In most countries, poverty rates are highest in remote rural areas, including forests. Forest dwellers and rural peoples still suffer from insecure and limited rights. In addition, economic growth in forest-rich developing countries lags behind that of developing countries with less than one third of their territories forested. What successes have forest dwellers and indigenous peoples seen from the past five decades of development investments and global proclamations?

Limited Recognition of Rights and Extensive Poverty

In much of the developing world, the human, civil, and political rights of forest-dwelling communities, including indigenous peoples, are denied or insecure. Governments in developing countries claim ownership and assert direct control over some 70 percent of the total area of forestlands, even though indigenous peoples, local communities, and households have legitimate, long-standing customary ownership of much—in many places, the majority—of these lands. Describing the situation, Liz Alden Wily writes:

> At the stroke of a pen, several billion people around the colonized world on four continents were rendered tenants of the state, with varying degrees of protection as mere occupants and users—not owners. Despite reforms, most remain so today... (RRI 2008, p. 9).

Forest and land laws commonly ignore, limit, or deny the rights of local communities and indigenous peoples in forests (Colchester et al. 2006). Even in countries where land rights are recognized, rights to use and benefit from forests are often constrained heavily by forest and land-use regulations (Scherr, White, and Kaimowitz 2004). This lack of recognition of the local rights of indigenous peoples living in forests tends to contradict international human rights laws—widely ratified by developing-country governments—that require the recognition of human, civil, and political rights. These include the right of indigenous peoples to own, use, control, and manage the lands and natural resources they customarily have occupied or used.

The mandates and programs of forest agencies, generally designed to generate financial revenues to government through commercial harvesting and to establish public protected areas, are often at odds with local people's human, civil, and political rights specified in national constitutions and land laws.

Poverty is disproportionately prevalent in dense forest areas, and often particularly severe and long-lasting there (Sunderlin, Dewi, and Puntodewo 2007). Many of the world's indigenous and ethnic minority communities inhabit forest areas. For instance, in India, 84 percent of tribal and ethnic minorities live in forest areas (Mehta and Shah 2003, p. 501). One reason why poverty rates are high in forest areas is that tenuous property rights and oppressive regulatory frameworks prevent customary owners from benefiting from their forest assets.

In addition, forest communities tend to lack political power and often the means to stand up to outside interests who wish to exploit their land. Other reasons for marginalization include the following:

1. The remote rural areas where forests are located have been relatively untouched by economic modernization.
2. Forests tend to be distant from markets, and the distance reduces income-earning opportunities and increases marketing costs.
3. Forest rents (especially timber) are difficult to capture without high levels of investment and infrastructure.
4. Forests have been a refuge for migrants, including people fleeing conflict.
5. Public investment rates tend to be low in the remote rural areas where forests are located (Sunderlin et al. 2008).

Similarly, the poorest people in many communities are unable to protect their interests against village elites, who can take advantage of insecure customary regimes to privatize commonly held resources and otherwise capture benefits.

Export-Oriented, Forest-Based Industry

As developing countries emerged from colonialism, governments were keen to establish large-scale industries, believing them to be fundamental building blocks of economic growth and trade. International development banks and development assistance agencies financed investment in large-scale forest industry, and they promoted an industrial model based on large-scale forest concessions and the export of logs and timber (Westoby 1987). By the early 1960s, most development institutions had active forestry portfolios providing loans for construction of sawmills, pulp mills, and other major industries. Most governments persisted with the economic production models established during the colonial period, maintaining control over forestlands and allocating forests to commercial concessions (Karsenty 2007; Oyono 2007).

Today, this model is well established in national policy and legal frameworks, and it continues to receive support from international financial institutions. In Central Africa alone, approximately 50 million hectares of forest are in industrial concessions (Karsenty 2007, pp. 8–18). But experience and research show that this model has failed to produce the equitable economic growth and development desired. In many cases, this industrial model has resulted in rampant human rights abuses, corruption, limited generation of local employment, and adverse impacts on the health and livelihoods of forest peoples living in concession areas (Counsell, Long, and Wilson 2007; World Bank 2006, 2007; Forest Trends 2006).

International institutions and governments have also promoted large-scale forest plantations as a complement to industrial concessions and, in some cases, as a response to deforestation. Worldwide subsidies for forest plantation development are comparatively small: roughly $2 billion per year for forest plantation, compared with $400 billion per year for agriculture (White, Bull, and Maginnis 2006, p. 15). However, these subsidies still exceed overseas development assistance in the forestry sector (White, Bull, and Maginnis 2006). These subsidies to the plantation industry can undermine the economic viability of natural forest management and the small-scale enterprises that depend on it, further weakening both the incentives to manage natural forests and the potential for natural forests to contribute to social and economic development.

Environmental Protection

Environmentalists from the global North frequently neglect to acknowledge that the people-less protected-area conservation model emerged in the United States *only* after several hundred years of epidemics, ethnic cleansing, and war against the indigenous population. Since that time, this model of conservation has been exported around the world, yet the issue of "conservation refugees" remains relatively invisible in popular dialogue. Both the number and size of protected areas in the World Database on Protected Areas have grown more than tenfold since 1962 (http://www.wdpa.org). The protected-area model was implemented with the worthy intention of conserving biodiversity; but, in application, it generally failed to recognize the rights or even existence of local people, constituting at its worst a direct land grab (Brechin et al. 2003; Geisler 2003).

Conservation models have evolved over the past 30 years, and conservationists are paying increased attention to the protection of biodiversity and ecological values in a broader landscape. But the new urgency for putting key biodiversity areas under some form of protection in the face of climate change risks fueling a new green land grab. The international development community and high-profile conservation agencies have set ambitious new targets for creating new and consolidating existing protected areas, with inadequate analysis of the rights issues and a poor understanding of the human-nature relationships that could be sustained by different ownership and management models.

Social and Participatory Forestry

In recognition that industrial development and environmental protection were providing few benefits for the poor and that forest degradation remained a serious problem, some international donors, nongovernmental organizations, and governments in the 1970s started to promote what was dubbed "social forestry." The term referred to a range of activities that promoted the greater involvement of people in the management of community forests, and the restoration of forests in and around agricultural landscapes. Except in a limited number of forests in which customary rights were clearly recognized, social forestry was initially considered suitable only where the forest resource had already become severely degraded.

Early projects were often driven by government agency targets and bureaucratic processes, with limited tailoring to local needs, conditions, or political realities. With time, deeper engagement with local people began to reveal the complexity of land and forest rights in the broader landscape. Foresters started to realize that vast numbers of rural people still lived in and around and claimed rights to natural forests; and social forestry expanded to include forest areas previously owned or managed by governments.

Several lessons can be drawn from the three-decade experiment with social, community, and participatory forestry. In almost all cases, it proved nearly impossible for these investments to reorient forest agencies to a more people-friendly approach. Nor did it lead to fundamental reforms of forest policy and property, even when social consensus was moving in that direction. Interventions often lacked good understanding of the broader market and policy context, resulting in many poor people investing their land and labor but being unable to benefit commercially. In most cases, advocates missed opportunities to scale up local innovations and to modify the taxes, policies, and regulations that were crippling local enterprise.

Market-Based Conservation

In the 1990s, a new set of instruments and approaches grounded in market incentives emerged to promote sustainable forestry. One of the most significant instruments to emerge was independent forest certification, a voluntary process by which the planning and implementation of on-the-ground forestry operations are audited by a qualified and independent third party against a predetermined standard. Ironically, industrial forest concessions and commercial plantations have been most favored by this development because of their larger scale; and forest certification has expanded disproportionately in temperate regions and well-governed countries.

Another intervention has been the promotion of payments and markets for ecosystem services (PES), such as carbon sequestration, biodiversity conservation, and water catchment systems. PES may provide better opportunities to serve the forest-dependent poor than do other conservation measures: Forests provide many services that could eventually find markets. There is new interest in avoided deforestation, and ecosystem service markets could be bundled together to achieve economies of scale. And there is a broad set of actors interested in investing. However, there also are great challenges: PES schemes are plagued by many of the same problems that have hindered earlier approaches to forest conservation and management. Without concerted and well-designed effort, the costs and lack of capacity to manage risks will present significant barriers for small producers and communities. If PES systems can scale up without undermining the lot of the rural poor, it will depend on the degree to which markets can be shaped to respect local rights and governance systems.

Lessons

Although these models and interventions have clearly brought gains to many forest areas, they have often entrenched institutional, political, and market structures that keep rural people poor and forest areas insecure. It is no small task to change the politics of control and the concentration of wealth that lie at the root of the challenge. Nevertheless, there are many examples of external interventions that have influenced domestic policies—from direct approaches such as participatory land mapping and facilitating legal action, to more indirect strategic approaches such as support for local research and organizations. These activities help build local capacity for more informed dialogue, and they open more political space for local voices.

It is not surprising that forest areas are characterized by social and political underdevelopment and injustice. Urban-based political, economic, and environmental elites have maintained official public ownership over forest areas and have exploited them for their own benefit. These external elites have used technically focused public forest agencies to implement national or global notions of the public good—overwhelming local rights and aspirations (Larson and Ribot 2007; Peluso 1992). Social, economic, and environmental development programs have often become impositions—treating forest areas as hinterlands to be exploited for the social and economic benefit of others, to be protected on another's behalf, or to provide environmental services on someone else's terms. For the most part, indigenous and nonindigenous forest communities alike have been unable to use forests to pursue their own development.

Many governments increasingly are open to strategic advice—not prescriptions—and information about how other governments are dealing with contentious tenure and policy reform issues. Overall, however, governments and donors have tended to careen from one crisis or the latest "panacea" to the next, finding it more difficult to muster the political will or organizational capacity to address the underlying institutional problems that led to underdevelopment in the first place.

Perhaps the most important finding from the past 50 years of development intervention in forest areas is about what was *not* done. No serious, substantial attempt was made to recognize and clarify property rights in forest areas or to empower forest communities to advance themselves economically or politically.

Past development assistance has also shown that trying to plan and organize optimal social and economic development structures from outside a target group is not only morally wrong, but also ineffective. Local communities must be enabled to identify and negotiate their options, and to become flexible and resilient in coping with unexpected change.

Reasons for Hope

Ironically, after centuries of subjugation and marginalization, forest dwellers and other rural peoples might hold in their hands the fate of the wider world. Because tropical forests remain one of the world's most important global carbon sinks, safeguarding and preserving existing forests and the broader agroforestry landscape will be an important part of the global climate change mitigation strategy. As evidence increasingly demonstrates, indigenous peoples and traditional forest communities are often better at preserving and protecting forest areas than are conventional systems of publicly protected areas (Bray, Merino-Pérez, and Barry 2005). Moreover, there are at least 370 million hectares of tropical forests managed and conserved by forest communities and indigenous peoples—at least as much forest area as is conserved in publicly protected areas (Molnar, Scherr, and Khare 2004, p. 10). There are many lessons to be learned from the experiences and knowledge of these forest communities. Recognizing and strengthening their ownership and management rights will support their continued practice of traditional management systems that have preserved their standing forests to date.

Moreover, as global climate change mitigation mechanisms take form, reducing deforestation and degradation of forestlands is emerging quickly as one of the most cost-efficient and effective means of mitigating carbon emissions. Forests and forest peoples are returning to the center stage as reducing emissions from deforestation and forest degradation (REDD) mechanism captures global attention as one option for inclusion in the post-2012 climate agreement.

Fortunately, among other key global trends, there is strong evidence that forest peoples are organizing themselves and gaining strength around the world. There has been a substantial growth in the number and capacity of indigenous peoples and community organizations; and despite often facing persecution, they are advancing their agendas for political and social development and for engaging in economic activities and enterprises. These trends strengthen the ability of forest dwellers and rural people to hold the rest of the world accountable for its actions.

Also encouragingly, some governments are beginning to rethink and rationalize property rights in forest areas by recognizing the territorial rights of local communities and indigenous peoples, and by attempting to clarify the property rights of households and individual citizens. About one half of all agrarian states—those countries whose economic structures are dominated by agriculture—have tenure reforms under way, including forest tenure in most of those countries (Alden Wily 2006, p. 26; Alden Wily and Mbaya 2001). Tanzania, for example, has led the way by establishing clear community ownership over land as the foundation for forest conservation and development, and thereby has influenced trends across Africa (Alden Wily 1998, 1999, 2001, 2002). Brazil and other countries in Latin America increasingly have recognized the territorial rights of indigenous peoples. In the past two years alone, new forest tenure policies or legislation have been adopted in Bolivia, Brazil, China, India, Indonesia, and the Russian Federation— affecting almost half the world's forest areas (Sunderlin, Hatcher, and Liddle 2008, p. 10; RRI 2008). Forest agencies increasingly accept the importance of secure property rights in putting the forest sector on a sound institutional footing, and the need for transparency to achieve effective public governance. Some governments are beginning to reverse historical obstacles to social inclusion by allowing rural people and civil society truly to participate in forest governance.

There has never been a greater opportunity to take advantage of this momentum and investment to help governments, communities, and private sector actors pursue equitable governance and development in forest areas. The next few decades are critical for addressing climate change and the underlying causes of social, political, and ecological insecurity that threaten forests and forest peoples.

Climate Change: Catastrophe and Opportunity

Both social and ecological systems will undergo major adjustments as a result of climate change, which already is affecting some of the poorest and most vulnerable communities around the world (Amazon Alliance/Forest Peoples Programme/RRI 2008; Roberts and Parks 2007; UNDP 2007). Poor people dependent on forest areas and other natural resources will be exposed and vulnerable to a wide range of changes to weather, rainfall, vegetation, and the distribution of wild fauna populations and migrations.

It is widely accepted that average mean temperatures will increase by at least 1–2 degrees Celsius. According to the *Stern Review,* this increase could cause the extinction of 15–40 percent of species; and add pressures that would force millions of people into extreme poverty, including those with limited and insecure rights to their lands, forest areas, and other natural assets (Stern 2006, p. 55).

Carbon Finance

The interlinked crises of climate change and energy are driving financial flows; land-use allocations; and a new international architecture of institutions, markets, and regulations. There are already pledges from major government donors for large investments in forests, such as Norway's September 2008 pledge to contribute up to $1 billion to reduce deforestation in the Brazilian Amazon. Emerging carbon markets will also drive tremendous investment in forests. By one estimate, reducing global deforestation rates by as little as 10 percent could generate between $2.2 billion and $13.5 billion in carbon finance annually (Ebeling and Yasué 2008, p. 1918).

What will these tremendous investments mean for indigenous peoples and poor forest-dependent communities? The dangers are clear. The money flowing from carbon markets is unconcerned with, and so far unhindered by regulation to protect, the rights and lands of forest dwellers. Carbon finance markets are galloping ahead, but "the primary goal of carbon financing is to offset emissions and not [to] guarantee pro-poor development" (Luttrell, Schreckenberg, and Peskett 2007, p. 2).

Mitigation and Adaptation Options in Forests

The United Nations Framework Convention on Climate Change (UNFCCC) articulates two approaches for addressing climate change: mitigation, or reducing emissions and increasing carbon sequestration; and adaptation, or adjusting to the changing climate. Forest management will play a key role in both approaches.

Forest management practices can be more sustainable when local communities are landowners, or at least have clear user rights (Brown, Chapin, and Brack 2006; Pokharel et al. 2006; Pierce and Capistrano 2005; RECOFTC 2004; Ribot and Larson 2004). This is not always the case. For example, a study in Ghana found that security of tenure was not an important factor in the practice of sustainable forestry (Owubah et al. 2001). Forests that are managed in a more sustainable manner are likely to be less vulnerable to climate change (Murdiyarso, Herawati, and Iskandar 2005; Robledo and Forner 2005; Reid et al. 2004). Thus, vulnerability to climate change can, under some circumstances, be reduced by the reform of forest tenure and user rights in favor of local communities.

The *Stern Review* concludes that "major institutional and policy challenges" would have to be overcome to realize the climate and social benefits of avoided deforestation, including clarifying forest-related property rights, strengthening law enforcement, and overcoming entrenched systems of vested interests. A more recent report concludes that "adaptation assistance needs to be integrated into development spending to deliver development goals in a climate resilient manner rather than being earmarked for climate-specific projects. This will require involvement of organizations and institutions beyond the UNFCCC" (Stern 2008, p. 36). To date, however, there is little evidence that institutional interventions to address climate change adaptation will consider forest tenure and user rights adequately.

Climate change mitigation proposals focused on forests concentrate on reducing greenhouse-gas emissions by reducing deforestation and forest degradation and by promoting afforestation. Many competing schemes (mostly devised by governments, conservation nongovernmental organizations, and the private sector) and funds (mostly promoted by the World Bank and donor governments) are on the table.

Adoption of these schemes backed by the necessary funding will have a significant impact on how forests are managed in coming decades and on who will manage them, with implications for millions of forest-dependent people and communities. Many risks are associated with carbon forestry (Griffiths 2007). These risks include

- renewed and even increased state and "expert" control over forests
- support for anti-people and exclusionary models of forest conservation
- violations of customary land and territorial rights
- unequal and abusive community contracts
- land speculation, land grabbing, and land conflicts (competing claims for compensation for avoiding deforestation).

The Challenge of Compensating Communities for Their Carbon

Past policies have addressed community ownership of forests and lands inadequately, and have ignored the rights of these communities to benefit from their carbon. And many countries have not begun to address the property rights issues surrounding carbon emissions and trade.

Mired in issues of national sovereignty, most proposed schemes for emissions reduction from forest areas overlook questions of equity, ownership, benefit sharing, and development outcomes. Under the Kyoto Protocol, there are some simplified mechanisms for small-scale afforestation/reforestation (A/R) projects, developed to enable communities to participate more fully in the Clean Development Mechanism (CDM) contained in the Kyoto agreement. However, even these A/R projects have proved to be largely out of reach for poor forest communities. High installation and transaction costs associated with project preparation; and the need for clear property rights to land, resources, and carbon have made it very difficult for poor rural communities to initiate and benefit from A/R projects under the CDM (Robledo et al. 2008).

Opportunities and Risks

The global and frightening nature of climate change will keep national governments focused on forest areas and forestry issues; and open to negotiating with civil society and forest communities, including indigenous people. There is trèmendous potential to make climate-related investments in a manner that strengthens local rights, reduces rural poverty, protects remaining natural forest areas, and restores degraded forest areas, while reducing greenhouse-gas emissions. Properly devised rights recognition programs and participatory forest projects could constitute a low-cost option for reducing emissions, sequestering additional carbon, and increasing adaptive capacity.

Conversely, an approach that attempts to extend public regulatory authority beyond protected areas in an effort to control land use and deforestation would be counterproductive. It would reverse the pattern of devolving forest management authority and increase the potential for conflict.

Within the UNFCCC, debate on proposed new forest-related mechanisms (including REDD and carbon markets) has only touched on issues related to local rights to forest resources, equity, governance, and legitimacy. But, because of the need for high standards of implementation, monitoring and evaluation, good governance, and equitable approaches are critical. Without them, future forest-related climate change initiatives will benefit only a few (primarily wealthy) elites, and will reinforce existing economic disparities.

Foundations for Effectiveness and Equity: A Framework for Climate Change Investments

Climate change and the global reckoning it inspires present both opportunities and risks for forests and forest peoples. How climate change mitigation and adaptation agreements and policy guidelines are structured will have direct implications for the rights, security, and livelihoods of indigenous peoples and forest communities around the world, especially in tropical forest countries.

All new interventions should be tempered with the knowledge that previous international interventions in forest areas have had limited effect on the loss and degradation of forests. Clear rights and equitable governance structures are crucial to achieving climate mitigation and adaptation goals through forest investments. Moreover, without clear rights and good governance, massive flows of funds and attention to forests will inadvertently undermine existing progress in defending human rights and supporting the livelihoods of forest dwellers.

Forest-based climate change interventions must be locally appropriate; and designing global and regional frameworks for supporting and monitoring these projects will not be simple. Alone, technical guidance and advice to specific governments and industries will not be sufficient to address the major social justice challenges facing forest peoples.

To ensure effective investments for climate change mitigation and adaptation in forest areas, we recommend that policy agreements and investment mechanisms incorporate and build on four mutually reinforcing and self-correcting foundations:

1. *Recognize rights*—establish an equitable legal and regulatory framework for land and resources.
2. *Monitor more than carbon*—create an information infrastructure that monitors more than carbon—perhaps rights violations and distribution of benefits—and is both transparent and easily accessible to the public.
3. *Provide independent advice and auditing*—establish and support independent advisory and auditing mechanisms at national and international levels.
4. *Pay the right people*—prioritize investments in communities and their forests.

Recognize Rights

Establish the legal and regulatory basis for effective climate change mitigation and adaptation efforts by increasing investments in recognizing and strengthening indigenous and other community rights to forestlands, trees, and their carbon in forest countries. First, governments and civil society actors must invest in recognizing and strengthening existing rights in forest areas—with particular attention to complex customary systems of ownership and management that do not translate neatly into existing legal frameworks. Doing so will require legal reforms to recognize the collective rights of indigenous and traditional forest communities and formally to recognize local peoples' rights to the ecosystem services (including carbon sequestration) their forest resources create. These investments should also include mapping and delimitation of forestlands, and strengthening the capacity of civil society to participate in and inform the design and implementation of forest tenure reforms.

Second, governments and civil society will need to build regulatory institutions to govern carbon rights and carbon markets. Both governments and civil society will need resources and support to monitor and learn from climate change mitigation and adaptation initiatives.

Monitor More than Carbon

Establish the information infrastructure for the independent and transparent monitoring of the status of forests, forest carbon, and the impacts on rights and livelihoods. Monitoring forest carbon alone will be insufficient to achieve effective climate change mitigation and adaptation, given the huge risks of inadvertent and negative effects on local rights and livelihoods. There will be considerable new investment to establish credible inventories of forests and forest carbon. Indeed, in many places, this work already has begun. These investments must be accompanied by parallel investment in clarifying and mapping ownership and access rights to forestlands.

First, governments and investors should establish credible maps and inventories of forest ownership and access rights. Doing so also will require investment in public knowledge and understanding of these rights, especially in forest communities and rural areas where governance and the rule of law are weak.

Second, governments and investors will need to monitor the status of payments, the distribution of benefits, and the local impacts of carbon payment structures on rights and livelihoods. These systems should monitor compliance with social safeguards and policies, especially local community rights to free, prior, and informed consent to activities in their communities and on their lands. These monitoring systems may emerge at multiple levels, with activities at local, national, and international levels. All monitoring must be transparent; and information should be easily accessible to the public.

Ensure Independent Advice and Auditing

Establish civil society advisory processes to guide, monitor, and audit investments and actions at national and global levels. Input from civil society and local voices will be essential to meeting climate change mitigation and adaptation goals equitably and efficiently. Technical guidance alone will not be enough, and neither will ad hoc social sector inputs to national and international policy and projects.

For independent advice and auditing to mitigate the risk that climate investments will exacerbate social tensions, two key steps will be required. First, there should be support to establish a civil society advisory group to formally advise the relevant bodies of the UNFCCC. This civil society group should monitor the design and implementation of investments in climate adaptation and mitigation in forest areas and policies at the global level. The group also could formally advise relevant groups within the international institutions, such as the UNFCCC REDD contact group, the UN-REDD initiative, and the World Bank Forest Carbon Partnership Facility.

Second, investors and policy makers should establish a fully transparent and locally led process of civil society consultation in forest countries. These national consultations could devise, guide, and monitor national forest reform and climate strategies, building on the lessons and experiences of the Voluntary Partnership Agreement negotiations conducted by the European Union and trade partner countries.

Pay the Right People

Ensure that national-level investments and climate investments in forest areas prioritize payments to indigenous peoples, forest communities, and rural peoples whose lives and livelihoods depend on and shape forests. Successful climate investment strategies must recognize the crucial historic role that forest dwellers play in maintaining and protecting forest cover by paying communities for avoided deforestation. Such payments will promote the stability of carbon sequestration in standing forests.

To achieve this, first, investments and climate action strategies must promote national adaptation plans that call for and work toward recognition and strengthening of local land and resource rights. National adaptation plans should include the full participation of forest peoples and rural communities in planning these strategies in locally appropriate ways.

Second, mitigation payments need to prioritize the traditional owners of natural forests and agroforests. Mitigation payments cannot reward past industrial clearing or degradation of forestlands.

To be effective and equitable, the majority of funds dedicated to adaptation and carbon emissions reduction from avoided deforestation should go to forest communities and households in forest areas.

Build a Framework

If established, those four foundations for climate investments would be mutually reinforcing and self-correcting over time. Inadequate effort or progress on one will weaken or undermine the other foundations, destabilizing the process and threatening opportunity for success.

These ideas are not new, and indeed projects are already under way to establish equitable payment schemes, address legal frameworks, and establish new monitoring mechanisms to support climate intervention goals. We need a more concerted effort to understand these mechanisms, and investment in a comprehensive approach that will work toward effective climate change investments that do not undermine recent progress in strengthening rights and social development in forest areas.

Conclusion

> There is a window of opportunity for avoiding the most damaging climate change impacts, but that window is closing: the world has less than a decade to change course. Actions taken—or not taken—in the years ahead will have a profound bearing on the future course of human development. The world lacks neither the financial resources nor the technological capabilities to act. What is missing is a sense of urgency, human solidarity and collective interest (UNDP 2007, http://hdr.undp.org/en/reports/global/hdr2007-2008/).

Windows of political opportunity for investment and reform can open suddenly and close just as quickly. It is in these moments of openness and opportunity that the global development community immediately should engage to help governments, civil society, and investors rethink and reform options, governance, and tenure in forest areas. Despite the challenges we face, the opportunity of this moment has never been greater.

In the next few decades, governments and investors will spend billions of dollars on energy, food, and climate change–related projects in or near forest areas. These projects will be effective and long-lasting only if they also are equitable. They will avoid exacerbating marginalization, tension, and conflict only if they help repair weak systems of governance and if they respect and support the rights of forest communities. The development record clearly shows that riding roughshod over local rights and local initiatives creates disparities in wealth that cannot be reconciled by further growth and investment, and a discontent that cannot be controlled by security forces.

Diversity is the key to adapting to climate change: there must be diversity in land-use systems, scales of production, local institutions, and cultural and social values. Small-scale enterprises and diverse agroecological, silvicultural, and pastoral systems will provide the greatest flexibility and resilience to rapid change and uncertainty. Rather than centralized mechanisms and top-down plans, we need open, responsive, and equitable processes of decision making that enable local people and regional governments to devise their own solutions to national and global challenges.

Thirty years ago, Jack Westoby saw the future for traditional development assistance models in forest areas:

> Wise governments will digest and apply the lessons of the last two decades of bitter experience. They will take a cool and calm look at the prospects of quick and easy export earnings...and not sign away their resource heritage.... The choice between need-oriented industry and profit-oriented industry is a political, not an economic choice. Once power is exercised by or on behalf of the broad population, then, and then only, will the contribution of forest industries to socio-economic development start to be realized (Westoby 1987, p. 247).

Terms like "human rights," "land reform," and "governance" may still be uncommon in the strategies and priorities of forest development and climate change experts. But there is compelling evidence that these concepts can become our most effective tools. There is much we can learn from our history and previous experience in other sectors. These lessons demonstrate that it is possible to carry out reforms, recognize rights, avoid deforestation, reduce conflict, heal divisions, and improve the livelihoods and well-being of communities. Most important, these lessons also show us that many millions of rural and forest peoples are anxious to move ahead. We just need to give them the chance.

References

Alden Wily, Liz. 1998. "The Legal and the Political in Modern Property Management: Re-making Communal Property in Sub-Saharan Africa, with Special Reference to Forest Commons in Tanzania." Presentation at "Crossing Boundaries," the Seventh Annual Conference of the International Association for the Study of Common Property, Vancouver, British Columbia, Canada, June 10–14.

———. 1999. "Community-Based Land Tenure Management: Questions and Answers about Tanzania's New Village Land Act." Issue Paper 120, International Institute for Environment and Development, London.

———. 2001. "Forest Management and Democracy in East and Southern Africa: Lessons from Tanzania." Gatekeeper Series 95, International Institute for Environment and Development, London.

———. 2002. "Community Forest Management in Africa: Progress and Challenges in the 21st Century." Proceedings of the 2nd International Workshop on

Participatory Forestry in Africa, Food and Agriculture Organization, Arusha, Tanzania, February 18–22.

———. 2006. "Land Rights Reform and Governance in Africa: How to Make It Work in the 21st Century?" United Nations Development Programme, New York.

———. 2008a. "Whose Land Is It? Commons and Conflict States: Why the Ownership of the Commons Matters in Making and Keeping Peace." Paper presented in panel on "The Contested Commons: From Conflict to Peace." International Association for the Study of the Commons Biannual Conference, Cheltenham, U.K., July 15.

———. 2008b. "Custom and Commonage in Africa: Rethinking the Orthodoxies." *Land Use Policy* 25 (1): 43–52.

Alden Wily, Liz, and Sue Mbaya. 2001. "Land, People and Forests in Eastern and Southern Africa at the Beginning of the 21st Century: The Impact of Land Relations on the Role of Communities and Forest Future." International Union for the Conservation of Nature, Nairobi, Kenya.

Amazon Alliance, Forest Peoples Programme, and RRI (Rights and Resources Initiative). 2008. "Community Forest Tenure, Governance and Benefits: The Missing Link to Climate Change Mitigation and Adaptation." Panel presentation at the International Union for the Conservation of Nature World Conservation Congress, Barcelona, Spain, October 8.

Bray, David Barton, Leticia Merino-Pérez, and Deborah Barry, eds. 2005. *The Community Forests of Mexico: Managing for Sustainable Landscapes*. Austin, TX: University of Texas Press.

Brechin, Steven R., Peter R. Wilshusen, Crystal L. Fortwangler, and Patrick C. West, eds. 2003. *Contested Nature: Promoting International Biodiversity with Social Justice in the Twenty-First Century*. Albany, NY: State University of New York Press.

Brown, David, Mac Chapin, and Duncan Brack. 2006. "Meeting 6: Rights and Natural Resources: Contradictions in Claiming Rights." In *Human Rights and Poverty Reduction: Realities, Controversies and Strategies,* ed. Tammie O'Neil, 77–84. London: Overseas Development Institute.

Colchester, Marcus. 2008. "Beyond Tenure: Rights-Based Approaches to Peoples and Forest Areas. Some Lessons from the Forest Peoples Programme." Forest Peoples Programme and Rights and Resources Initiative, Washington, DC. http://www.rightsandresources.org/documents/files/doc_825.pdf.

Colchester, Marcus, Marco Boscolo, Arnoldo Contreras-Hermosilla, Filippo Del Gatto, Jessica Dempsey, Guillaume Lescuyer, Krystof Obidzinski, Denis Pommier, Michael Richards, Sulaiman N. Sembiring, Luca Tacconi, Maria Teresa Vargas Rios, and Adrian Wells. 2006. *Justice in the Forest: Rural Livelihoods and Forest Law Enforcement*. Forest Perspectives 3. Bogor Barat, Indonesia: Center for International Forestry Research.

Counsell, Simon, Cath Long, and Stuart Wilson, eds. 2007. *Concessions to Poverty: The Environmental, Social and Economic Impacts of Industrial Logging Concessions in Africa's Rainforest Areas.* London: Rainforest Foundation and Forests Monitor.

Ebeling, Johannes, and Maï Yasué. 2008. "Generating Carbon Finance through Avoided Deforestation and Its Potential to Create Climatic, Conservation and Human Development Benefits." *Philosophical Transactions of the Royal Society B: Biological Sciences* 363 (1498): 1917–24.

FAO (Food and Agriculture Organization of the United Nations). 2006. "Better Forestry, Less Poverty." Forestry Paper 149, FAO, Rome.

Forest Trends. 2006. "Logging, Legality and Livelihoods in Papua New Guinea: Synthesis of Official Assessments of the Large-Scale Logging Industry." Volume II. Washington DC: Forest Trends.

Geisler, Charles C. 2003. "Your Park, My Poverty: Using Impact Assessment to Counter the Displacement Effects of Environmental Greenlining." In *Contested Nature: Promoting International Biodiversity with Social Justice in the Twenty-First Century,* ed. Steven R. Brechin, Peter R. Wilshusen, Crystal L. Fortwangler, and Patrick C. West, 217–29. Albany, NY: State University of New York Press.

Griffiths, Tom. 2007. "Seeing 'RED'? 'Avoided Deforestation' and the Rights of Indigenous Peoples and Local Communities." Forest Peoples Programme, Moreton-in-Marsh, U.K.

Kaimowitz, David. 2005. "Forest Areas and Conflicts." *European Tropical Forest Research Network Newsletter* 43/44: 5–7.

Karsenty, Alain. 2007. "Overview of Industrial Forest Concessions and Concession-Based Industry in Central and West Africa and Considerations of Alternatives." French Agricultural Centre for International Development (CIRAD), Paris.

Larson, Anne M., and Jesse C. Ribot. 2007. "The Poverty of Forest Policy: Double Standards on an Uneven Playing Field." *Sustainability Science* 2 (2): 189–204.

Luttrell, Cecilia, Kate Schreckenberg, and Leo Peskett. 2007. "The Implications of Carbon Financing for Pro-Poor Community Forestry." *Forestry Briefing* 14: 1–5. London: Overseas Development Institute.

Mehta, Aasha Kapur, and Amita Shah. 2003. "Chronic Poverty in India: Incidence, Causes and Policies." *World Development* 31 (3): 491–511.

Menzies, Nicholas K. 2007. *Our Forest, Your Ecosystem, Their Timber: Communities, Conservation, and the State in Community-Based Forest Management.* New York: Columbia University Press.

Molnar, Augusta, Sara J. Scherr, and Arvind Khare. 2004. "Who Conserves the World's Forests? Community-Driven Strategies to Protect Forests and Respect Rights." Policy Brief, Forest Trends and Ecoagriculture Partners, Washington, DC.

Murdiyarso, Daniel, Hety Herawati, and Haris Iskandar. 2005. *Carbon Seques-tration and Sustainable Livelihoods: A Workshop Synthesis.* Bogor Barat, Indonesia: Center for International Forestry Research.

Owubah, Charles E., Dennis C. Le Master, J. Michael Bowker, and John G. Lee. 2001. "Forest Tenure Systems and Sustainable Forest Management: The Case of Ghana." *Forest Ecology and Management* 149 (1/3): 253–64.

Oyono, Philip René. 2007. "Understanding Forest Tenure in Central Africa: Transitions or Hidden Status Quo at the Dawn of the New Century?" Background paper for the Listening, Learning and Sharing Launch of the Rights and Resources Initiative, International Union for the Conservation of Nature and Rights and Resources Initiative.

Peluso, Nancy Lee. 1992. *Rich Forests, Poor People: Resource Control and Resis-tance in Java.* Berkeley, CA: University of California Press.

Pierce, Carol J., and Doris Capistrano. 2005. *The Politics of Decentralization: For-ests, Power and People.* London: Earthscan.

Pokharel, Bharat, Dinesh Paudel, Peter Branney, Dil Bahadur Khatri, and Mike Nurse. 2006. "Reconstructing the Concept of Forest-Based Enterprise Develop-ment in Nepal: Towards a Pro-Poor Approach." *Journal of Forest and Liveli-hood* 5 (1): 53–65.

RECOFTC (Regional Community Forestry Training Center for Asia and the Pacific). 2004. "Community Forestry and Good Governance Initiatives in Asia: Exploring the Synergies." Bangkok, Thailand.

Reid, Hannah, Saleemul Huq, Aino Inkinen, James MacGregor, Duncan Mac-queen, James Mayers, Laurel Murray, and Richard Tipper. 2004. "Using Wood Products to Mitigate Climate Change: A Review of Evidence and Key Issues for Sustainable Development." International Institute for Environment and Devel-opment, London; and Edinburgh Centre for Carbon Management, Edinburgh, Scotland.

Ribot, Jesse C., and Anne Larson, eds. 2004. *Decentralization of Natural Resources: Experiences in Africa, Asia and Latin America.* London: Routledge.

Roberts, J. Timmons, and Bradley C. Parks. 2007. *A Climate of Injustice: Global Inequality, North-South Politics, and Climate Policy.* Cambridge, MA: MIT Press.

Robledo, Carmenza, Jürgen Blaser, Sarah Byrne, and Kaspar Schmidt. 2008. "Climate Change and Governance in the Forest Sector: An Overview of the Issues on Forests and Climate Change with Specific Consideration of Sector Governance, Tenure, and Access for Local Stakeholders." Rights and Resources Initiative, Washington, DC.

Robledo, Carmenza, and Claudio Forner. 2005. "Adaptation of Forest Ecosystems and the Forest Sector to Climate Change." Food and Agriculture Organization of the United Nations, Rome; and Swiss Agency for Development and Coopera-tion, and Intercooperation, Berne, Switzerland.

RRI (Rights and Resources Initiative). 2008. *Seeing People Through the Trees: Scaling Up Efforts to Advance Rights and Address Poverty, Conflict and Climate Change.* Washington, DC: RRI.

Scherr, Sara J., Andy White, and David Kaimowitz. 2004. "A New Agenda for Forest Conservation and Poverty Reduction: Making Forest Markets Work for Low-Income Producers." Forest Trends, Washington, DC.

Stern, Nicholas. 2006. *The Economics of Climate Change: The Stern Review.* Cambridge, U.K.: Cambridge University Press.

———. 2008. "Key Elements of a Global Deal on Climate Change." London School of Economics and Political Science, London.

Sunderlin, William D. 2007. "Poverty, Rights and Tenure on Forest Lands: The Problem, and Priority Actions for Achieving Solutions." Working Paper, Rights and Resources Initiative, Washington, DC.

Sunderlin, William D., Sonya Dewi, and Atie Puntodewo. 2007. *Poverty and Forests: Multi-Country Analysis of Spatial Association and Proposed Policy Solutions.* Occasional Paper 47. Bogor Barat, Indonesia: Center for International Forestry Research.

Sunderlin, William D., Sonya Dewi, Atie Puntodewo, Daniel Miller, Arild Angelsen, and Michael Epprecht. 2008. "Why Forests Are Important for Global Poverty Alleviation: A Spatial Explanation." *Ecology and Society* 13 (2): 24.

Sunderlin, William D., Jeffrey Hatcher, and Megan Liddle. 2008. "From Exclusion to Ownership? Challenges and Opportunities in Advancing Forest Tenure Reform." Rights and Resources Initiative, Washington, DC.

UNDP (United Nations Development Programme). 2007. *Human Development Report 2007/2008. Fighting Climate Change: Human Solidarity in a Divided World.* New York: UNDP.

Vidal, John. 2008. "The Great Green Land Grab." *The Guardian* February 13: 6. http://www.guardian.co.uk/environment/2008/feb/13/conservation.

West, Paige, James Igoe, and Dan Brockington. 2006. "Parks and Peoples: The Social Impact of Protected Areas." *Annual Review of Anthropology* 35: 251–77.

Westoby, Jack. 1987. *The Purpose of Forests: Follies of Development.* New York: Basil Blackwell.

White, Andy, Gary Q. Bull, and Stewart Maginnis. 2006. "Subsidies for Industrial Plantations: Turning Controversy into Opportunity." *Arborvitae* 31: 15.

World Bank. 2006. "Inspection Panel Investigation Report. Cambodia: Forest Concession Management and Control Pilot Project." World Bank, Washington, DC.

———. 2007. "Final Report of World Bank Inspection Panel Investigation of Forest Sector Operation in DRC." Report 40746-ZR, World Bank, Washington, DC. http://siteresources.worldbank.org/EXTINSPECTIONPANEL/Resources/FINALINVREPwhole.pdf.

Figures, notes, and tables are indicated by f, n, and t following page numbers.